Out of the Field:

International Harvester Refrigeration

BY
SARAH TOMAC
COPYRIGHT 2023
1831 PRESS

IBSN 978-1-952265-03-7
International Harvester Refrigeration
        Out of the Field Series, Book II

All images in this publication with the WHS Image identification number in caption are used with
the written permission of the Wisconsin Historical Society, wisconsinhistory.org, in conjunction
with Navistar, inc. (a member of the TRATON Group)

All other images in this publication are from private collections of original literature from
International Harvester Company of America and International Harvester Company of Australia
and International Harvester Company of Canada.
Special Thanks to:
Gregg Montgomery, former International Harvester Design Manager, for his concept drawings.
Travis and Meghan Loschen Family for all their help with the specifics of the models and more
than I can list. (Irma Harding will get her own book, this is just beginning.)

Matt Weessies. for the HVAC lessons, the encouragement, the countless bowls of ice cream and
pies, and finding more reference materials than I remember having borrowed from you.

For the Spirit of Harvester
For the Prosperity of Agriculture
For the Faith that has Shown
For you.

OUT OF THE FIELD:
INTERNATIONAL HARVESTER REFRIGERATION

Beginning Development

Refrigeration.

Something we all rely on almost daily – it can be found in many sources today, in everyday use, like our refrigerator, car air conditioning, walk-in coolers, freezers, cooling tanks, dehumidifiers, and even ice rinks.

The purpose of refrigeration is to remove heat. In early times, before power sources became readily available, generations past relied on the coolness of flowing water or in the colder climates, ice.

Ice houses were common in many places, large warehouses could be found in major cities, and the Iceman made deliveries alongside the milkman. The first record of home delivery in America of natural ice in 1802 brings images of the delivery wagons of all types being driven around the streets. This influx of delivery led to many types of entrepreneurial endeavors, including the first commercial shipment of ice to Martinique from Boston in 1805, to help stop yellow fever.

Ice made by nature fluctuates, as seasons change, temperatures rise and fall, and the harvesting of ice is more successful in years than others. People, by design, like to investigate and have consistency. The advent of manufactured ice through science and experimentation has changed the way we store food today. Names like Faraday, Carre', Cullen, Gorrie, Perkins, Linde and many more are overlooked in the modern era of refrigeration, all of them with notable contributions to the modern method of refrigeration over the years.

Faraday, working in the Royal Institution of London around 1824, succeeded in condensing ammonia gas to a liquid by applying pressure and cooling it. Removing the pressure, the liquid boiled off as a gas, absorbing heat to do so. Any liquid absorbs heat when it turns to gas.

Left: Consistent cold water from a spring was one way to keep food cool. Above: Ice was often shipped long distances in the warmer states where snow and ice do not form reliably. Tomac Collection (TC)

In 1755, Dr. William Cullen invented the first machine to make ice by mechanical means, which was limited to cold temperatures to operate. This was improved upon by Carre' in 1858, using a solution of ammonia and water as the refrigerant to express cool water to form ice.

The first Carre' machine in the United States arrived via New Orleans, in 1863, despite the blockade during the Civil War preventing imports. This machine had a capacity of making 500 pounds of ice a day. Three more machines were purchased, and all of them were not successful, partly due to human operation error. These were later sold to a company in San Antonio, Texas and put in operation successfully. Jacob Perkins, an American living in England, in 1834, patented an early version of the modern compressor using ether as the refrigerant.

Dr. John Gorrie, in Appalachicola, Florida, registered the first patent in 1850 for a practical way of making ice.

The first successful ammonia compression machine was introduced between 1873 and 1875 by CPG Linde, of Germany and David Boyle of the United States. Over the following 15 years, various forms were introduced, and many improvements were made.

The general household relied on the manufacture of Ice to keep the contents of the icebox cold. Ice cools as it melts, absorbing the heat from the food to do so. The cool air circulates throughout the compartment during the process of removing the heat from the newly placed food, rising to the block of ice, and then the newly cooled air drops down, over the food and the process is repeated.

7,000 factories supplied America in 1933 with over 60 million tons of manufactured ice, as well as 7.5 million tons of natural ice every year. New York city used 4 million tons of ice on average each year and held a reserve of 250,000 tons for fluctuations. Daily production of factories averaged 23,000 – 24,000 tons. Many towns and cities with ice factories had similar reserves, or cold storage buildings, for unusually high demands of ice. New York city infrastructure required as many as one thousand trucks and wagons for delivery, employed about two thousand men, with a population of nearly seven million people.

Throughout the United States, you can still find a cold storage building, one in Kansas City has converted a cold storage building into lofts, a project that began in 2007, keeping the historical aspect of the original building built in 1922 in the River Market District.

The days of ice delivery have come to an end, and very few people alive one hundred years later remember the winter chore of harvesting ice for cool refreshing drinks in July or August. By the mid-1930s, refrigeration was being used in many industrial areas, mainly in refrigerated boxcars for transport of foodstuffs, mostly for meat processors and dairy industries.

A man is unloading a block of ice from the back of a D-30 in 1937. Delivering ice was common until late 1940s. WHS 88843

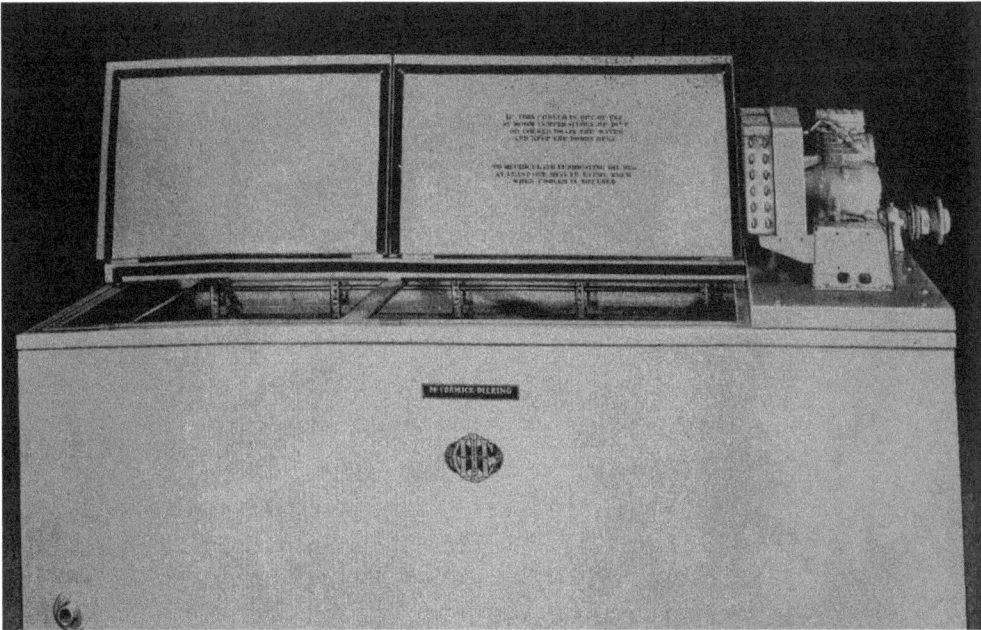

Number 10 Milk Cooler, wet type, as pictured in a milk cooler sales catalog about 1938. This cooler is designed to be powered by an engine or line shaft operation. Handle these coolers with caution, asbestos was used as insulation before fiberglass became common. TC.

When freon was discovered in 1928, it became a much more affordable option to use as a coolant for refrigeration. Production of cooling units, air conditioners, and ice making machines were limited to a handful of large companies with the financial means to produce these units prior to the discovery. Cooling systems require a gas of some type to act as the refrigerant, which absorbs the heat inside the unit. The lack of heat is what makes items cold.

Many different gases were used in the beginning of refrigeration development.

In 1933, the most common refrigerant used was Sulphur Dioxide. Other refrigerants used were Methyl Chloride, ethyl chloride, butane, isobutane, ammonia, propane, carbon dioxide and ether.

Ammonia was the preferred gas to use in large commercial plants, largely because of high latent heat, appealing working pressures and ease of detecting leaks. This was not the case for smaller household units, although the domestic fridges built in Australia used Ammonia.

Refrigerants, considered safe in 1933, that were non-inflammable, non-toxic and free from obnoxious odors, were used in areas where fire hazards and danger of odor exposure to large numbers of people, such as ships and air conditioning in theaters. The gasses used extensively for this purpose were Carbon Dioxide, diethylene, and methylene chloride.

Ether found use in small, hand operated units, mostly manufactured in Europe, and sold in the Tropics. Nitrous Oxide was used in chemical industries which required very low temperatures.

This man is removing a milk can from a 6 or 8 can cooler, with and electric driven motor. Notice the broken edges of concrete, indicating this was possibly a water tank for cooling milk cans before being converted for use with the electric milk cooler, in this old building. WHS 59597.

Sulphur dioxide and methyl chloride were preferred mainly because the atmospheric pressures were nearly the same, reducing the probability of leaks in the system. Gasses that require a very high or very low condensing pressure posed problems in either escaping or drawing air into the unit. Sealing of the tubing and joining to the compressor is very important to prevent loss of the refrigerant.

Dichlorodifluoromethane, a fluoride refrigerant, commonly known as freon or R-12, was gaining popularity as the latent heat of vaporization was both low and required lower pressurization of the systems as well. It also proved less corrosive than many other refrigerants, especially Sulphur dioxide, which when mixed with water, creates Sulphureous Acid, and becomes corrosive.

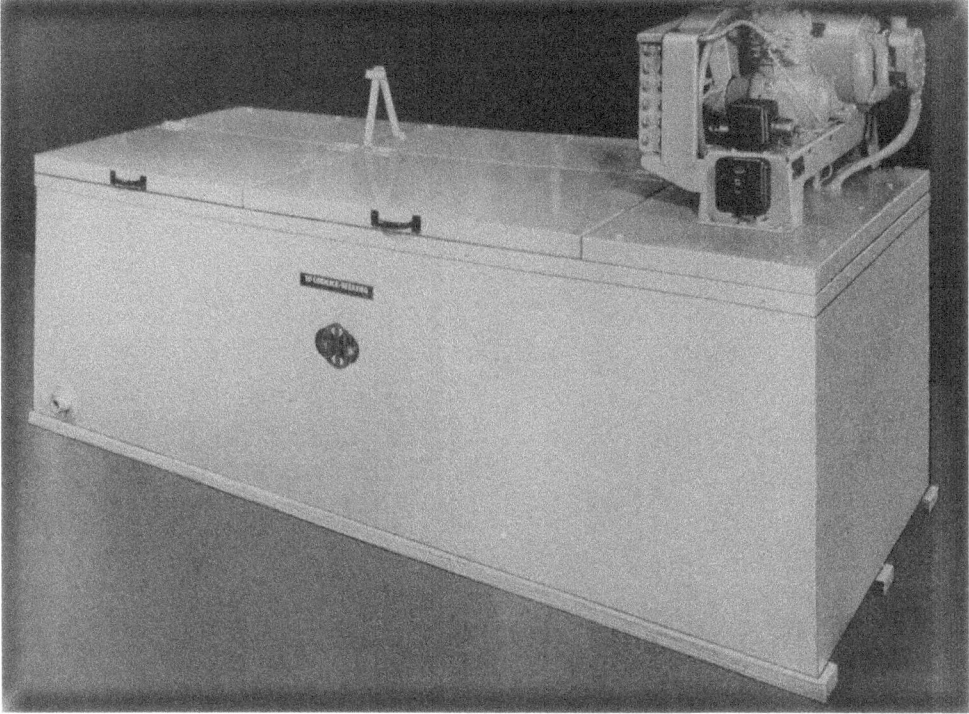

12 Can Electric driven Milk cooler, about 1938, as presented in a McCormick-Deering Milk Cooler catalog. TC .

Milk Coolers

Food Preservation has always been important, directly, and indirectly resulting in inventions that are life changing and surprising. Without the ability to freeze or preserve foods through cooling, we wouldn't be able to enjoy flash frozen fish, any vegetable, some fruits, or frozen dinner meals beyond the winter months in the coldest parts of the world. The advances in food storage with the ability to flash freeze would fill another chapter. This book is about refrigeration, not food. I think. Then again, I couldn't enjoy my nightly bowl of ice cream without the means of refrigeration.

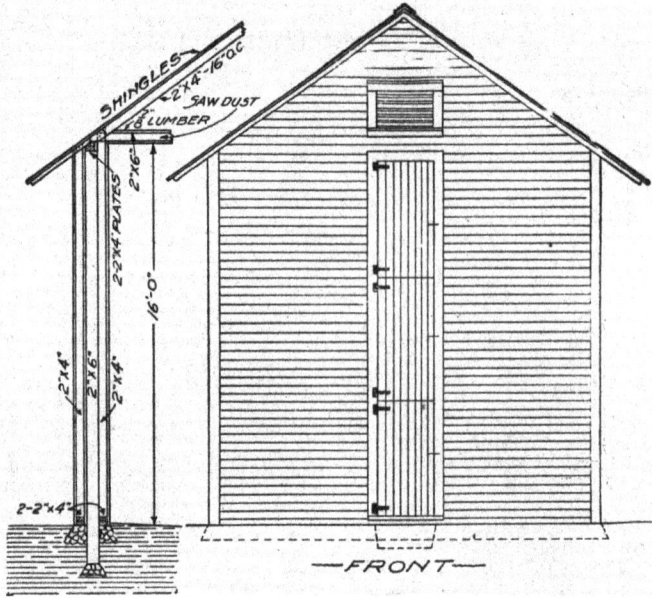

Drawing of Ice House as presented in the 1912 Osborne Almanac.
The almanac was printed by International Harvester for three years. TC.

The dairy industry benefited the most in the early development of refrigeration. Before the ability to mass cool, farmers had to know just how much ice to store on the farm to keep milk cold. The United States Department of Agriculture, in farmers' bulletin number 1078 titled "Harvesting and Storage of ice on the farm" covers, among other items concerning ice, how much is needed for the dairy herd. Location, number of cows, and methods of handling the product all add variants to the amount of ice needed to be stored. The consensus from the USDA in 1933 was that a half-ton of ice, or one thousand pounds, per cow milked was needed for cream and increase the amount to 1 ½ tons of ice in the North (cooler temperatures) and 2 tons per cow in the South.

According to the USDA guide, an average farm family needed five tons of ice for general use, one and a half tons per milking cow, and to increase the estimated amount by 20 percent to account for shrinkage and loss. A farm in the 1930s averaged about 12 cows, needing 18 tons of ice. One ton of ice is about three cubic feet, or 1.3 cubic yards. To store 26 tons of ice, a farmer needed to build an icehouse that could store about 34 cubic feet of ice. The icehouse interior would be almost 12' square, and just as tall. That's a lot of ice. Every year. It is easy to understand why farmers were quick to embrace the invention of refrigeration!

Milk storage in the 1930s met with strict dairy requirements. Farmers were required to be able to cool the milk uniformly and quickly to prevent bacteria growth. The cooling temperatures had to be met within an hour of milking, bringing the temperature down from the animals' body heat of 98°F to below 40°F. This prevents the growth of bacteria and protects the milk from spoilage. The days of using cold, fresh spring water had come to an end.

Two men are placing cooling coils and coil frame into a 6-can cooler as part of the assembly line at the International Harvester West Pullman Works in 1935. WHS 54791.

Not long after the awareness of the amount of ice to store for safe milk handling, International Harvester introduced in 1935 a milk cooler. These wet-type coolers were the first step to refrigeration as we are familiar with today.

Production of milk coolers began at the West Pullman Works in 1935, bringing the line of farm equipment into the field of farm refrigeration. West Pullman, Illinois, now a suburb of Chicago, was home to the Plano Manufacturing Company, which began in 1893. This company was one of the five that combined to form International Harvester Company on August 12, 1902.

The other four companies were McCormick Harvesting Machine Company, Deering Harvester Company, Milwaukee Harvester Company and Warder, Bushnell and Glessner Company.

The West Pullman Works produced a wide variety of items, from its beginning as the Plano grain binders, chains, mowers, wagons, and hay rakes. Later the 8-16 tractor, spreaders, planters, threshers, magnetos, and milk coolers were built.

The factory operated as a foundry and forge shop as well. In the end of the production time at Harvester, it was considered the precision products division, producing parts and assemblies for the other plants, mainly hydraulics, bearings, fasteners, and screw machine parts. The fifty-two-acre plant with 981,000 square feet of floor space over 35 buildings is now a solar park in 2023.

An inside view of a 6-can milk cooler, note the agitation of ice-cold water around the cans, assuring quick cooling of milk at low cost.

Below: Front view of 6-can electric driven Milk Cooler. TC

International Harvester, in 1934, introduced a wet type of milk cooler to help cool milk cans faster. This was the beginning of the early milk cooler or refrigerator. The water was agitated and circulated around inside the cooler to cool the milk cans faster. The gas used as refrigerant in the milk coolers in the beginning was Sulphur Dioxide. Powered by electric or an LA engine from one horse to 2 hp, this was a great way to keep milk cold on the farm for the milkman.

An insulated box, that could hold two to eight milk cans, with a hinged lid, these watertight steel boxes were built in the West Pullman works with the engines built in the Milwaukee Works. The last complete milk cooler built at the West Pullman Works was in June of 1946, serial number 33867.

Wet storage of tanks transitioned quickly to dry style coolers and used for more than beverages in watertight containers. The possibility of food storage increased by untold options. This was the first real step towards household refrigeration. Until this development, the cooling industry was focused on large, commercial prospects. Stores, bars, soda and drink retailers all could see the impact these smaller coolers were making, being able to sell icy cold drinks consistently.

 The very first wet-type cooler was bought for a dairy business in Dallas Center, Iowa by Mr. and Mrs. J D Keller. Not long after they bought the first cooler for the dairy operation, like many farmers, it was put into additional use. The Keller's also

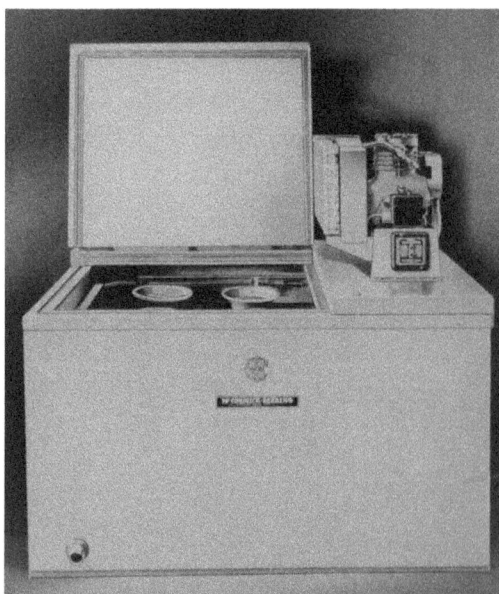

raised meat chickens and provided the dressed bird to local hotels and restaurants. On one occasion, the order was large and there was a time constraint that the family was working with. They found, by packing the birds into milk cans and using the cooler, the birds cooled faster and stayed fresh, allowing the Keller's to deliver the large order on time.

Many farms found the same uses and would order the units and use them as a large cooler or taking the cooling component from the milk cooler and turning it into a large cool room.

West Pullman Works managers standing with a refrigerator produced for the
U.S. Marines and delivered to the Philadelphia Navy Yard, March 19, 1941.
Mr. Watling, left, superintendent of West Pullman Works and Mr. Johnson, right,
Assistant superintendent, standing behind the unit.   WHS 59499

An option for the wet type agitation cooler, a dry box was available that could be inserted into the unit. This waterproof box was designed to lock into place on one side and be removed for cleaning of milk can cooling use. Development for a half dry and half wet cooler followed soon after as well as a fully dry type unit. These coolers were anywhere from four to ten cans in size. With the dry cooler, many farms quickly realized that you could store butter, meat and vegetables, ready for market.
Not long after the wet cooler was introduced, a whole industry from bakers to roadside stands were using the coolers for applications that Harvester did not even imagine at the time. Units were being used to cool piped drinking water, soda, or beer at roadside stands, bars and diners. Bakers, florists, photographers, mortuaries, construction and logging camps and milk stores were a few places that had sent letters and support of the use of the unit.
This type of loyalty came from the wide network of Harvester dealers and their reputation of providing a quality and reliable product.
The US military also found use for the cooler and ordered a self-contained no.12 unit with an International 1.5hp condensing unit powered by an LA 1.5-2.5HP Engine.
This cooler -turned-refrigerator was used to hold about one ton of meat, for use at a mess hall.

By 1940, there were 23 general types of units, in a variety of sizes from a small two cans size cooler to a large walk-in cooler. The milk cooler had become more than just a way to cool milk rapidly for the dairy industry.

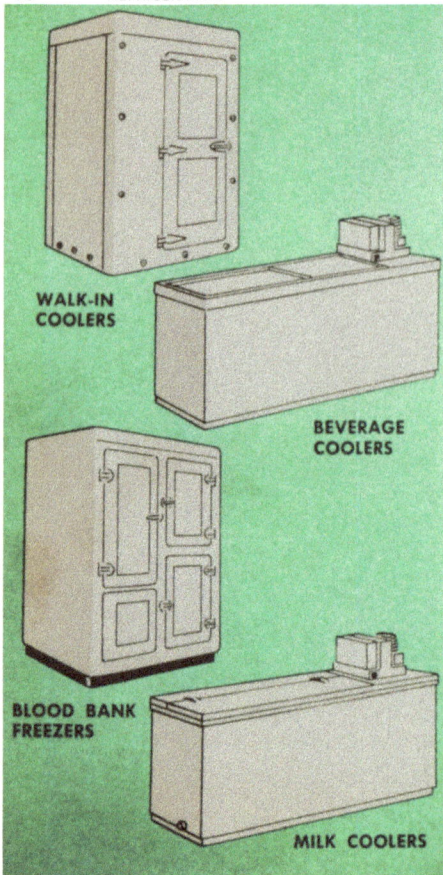

An assortment of the types of cooling equipment that International Harvester offered by 1940. Walk-in Coolers in a variety of sizes; Beverage coolers for various establishments; Blood Bank freezers for the medical professions and milk coolers, both wet type and dry type (no water for agitation)
Some of the walk-in coolers were ordered with a special feature to cool milk over an aerator by pumping it through chilled water, via copper tubing and a water tank, shown through open door. TC.

Part of this success was the widespread expansion of electric power to farms with the Rural Electrification Act of 1936. This Act was passed to ensure that all residences in the USA would have access to power. At the end of 1935, about 800,000 farms were electrified, and by 1940 more than 1,800,000 farms (nearly 27% of the nation's farms) were receiving power from a central station service. As more farms were able to access power, the demand for the milk cooler increased, allowing more dairy farms to meet the strict requirements for milk safety.

A man is removing a pan from a International Harvester Reach-in Cooler at Johnson's Bakery, 1941. Note the 'Kill' stamp on the photo, used to indicate that this image should not be used in promotions or any publication. WHS 60074

Reach-in and Walk-in

The components used for the dry cooler were quickly made available by Harvester for large walk-in coolers, for on-farm market use beginning in 1940.

The walk-in units were available through Harvester with many sizes available.

1940, the common refrigerant used was Methyl Chloride. Or Chloromethane, (R-40). Safer than the Sulphur dioxide that was previously used in the beginning, according to a sales manager promo in Harvester World, May 1940. Both of these are no longer used in the refrigeration business, as it has proven to be unsafe for that use.

Modern-day, when we require food that isn't grown by us, we venture out to a grocery store, wander amongst the aisles selecting the items needed for ourselves. This style of shopping was not common until the 1950s, many farms sold much like a farm market does today, direct, farm to fork.

Harvester soon found that the farmer knew how to adapt and utilize the product on offer and kept up with the feedback by creating an appealing package for everyone. Meat markets, vegetables, bakers all found ways to make use of this innovative design.

*A woman, probably a nurse or volunteer, places a container into a "Tomac Plasma Bank",
a refrigeration unit developed by International Harvester for the Tomac Medical Group,
1941. WHS 63570*

Reach-in coolers or powered ice boxes were common in the general consumer population. Harvester was commissioned by the Tomac Medical Group to build a blood
bank cooler, for the storage of plasma blood that was part of the war effort.

1944 – WF Borgerd, refrigeration engineer, directed the experimental and develop-
mental work for the refrigeration test department. He also directed the same devel-
opmental work that produced the Tomac Blood Plasma Bank, also manufactured by
the IH company.

The reach-in cooler was built and available as a natural progression from the dry-
type cooler, utilizing the same design, wrapping the Methyl Chloride lines around
the outside of the heavily insulated box, and placing the cooling unit at the bottom
corner on the smaller units and the cooling unit on the top for the larger ones.

Mrs. Wilbur Shrieder stands in a 6 x 12 foot refrigerated walk-in storage area, with a large Reach-In unit behind her, also holding a variety of meats and other items. WHS 60071.

Many of these reach-in coolers were compatible in size to multi-door commercial refrigerators today. The smallest at 30 cubic feet in volume measured 5' wide, 6 ½' tall and 2 ½' deep; while the largest was 7' wide, 3' deep and 6 ½' tall or 70 cu. ft. Some options you could hang a half hog or a quarter beef in one side, ready for the butcher to make the cuts needed at his place of business.

INTERNATIONAL HARVESTER MILK COOLER WARRANTY

International Harvester Company will service or replace free of charge (with new or rebuilt parts) all parts of the entire International Harvester Milk Cooler, identified by the model number and cabinet serial number on the reverse side hereof, if shown before ONE YEAR after date of delivery to the original purchaser to have been defective in material or workmanship when delivered.

Furthermore, it will service or replace free of charge (with new or rebuilt parts) the sealed-in refrigerating mechanism of said International Harvester Milk Cooler, exclusive of electrical controls, if shown before FIVE YEARS after date of delivery to the original purchaser that it was defective in materials or workmanship when delivered.

This warranty is in lieu of all other warranties expressed or implied.

INTERNATIONAL HARVESTER COMPANY

INTERNATIONAL HARVESTER FREEZER WARRANTY

International Harvester Company will repair or replace free of charge all parts of the entire International Harvester Freezer, identified by the serial number on the reverse side hereof, if shown before ONE YEAR after date of delivery to the original purchaser to have been defective in material or workmanship when delivered.

Furthermore, it will replace free of charge the sealed-in-refrigerating mechanism of said International Harvester Freezer, exclusive of electrical controls, if shown before FIVE YEARS after date of delivery to the original purchaser that it was defective in materials or workmanship when delivered.

This Warranty shall apply only within the boundaries of the Continental United States and the Territories of Alaska and Hawaii and is in lieu of all other warranties expressed or implied.

INTERNATIONAL HARVESTER COMPANY

> *The early mascot for the refrigeration department featured two polar bears on either side of the classic IHC logo and written on a curve around the logo is "International Refrigeration" in red, and a red circle defining this with a yellow background.*
>
> *The Polar bears impressed that the company made items worthy of the cold environment that they preferred. This logo was used in the beginning of the department, while the milk coolers and walk-in coolers were built. By 1947, with the introduction of the first freezer at the National 4-H Congress, this logo was no longer in use.*

The Warranty issued for the Milk coolers, Freezers and Household Refrigerator was a continuation of Harvester's commitment to the customer, offering warranties for their product since the beginning of building reapers in Chicago.

Cyrus was an innovator, and wanted all customers to be confident in purchasing the products, offering replacements for products that were a result of manufacturing failures, not the customer. This warranty was an industry first, and changed product support still today.

INTERNATIONAL HARVESTER HOUSEHOLD REFRIGERATOR WARRANTY

International Harvester Company will repair or replace free of charge all parts of the entire International Harvester Household Refrigerator, identified by the serial number on the reverse side hereof, if shown before ONE YEAR after date of delivery to the original purchaser to have been defective in material or workmanship when delivered.

Furthermore, it will replace free of charge the sealed-in refrigerating mechanism of said International Harvester Household Refrigerator, exclusive of electrical controls, if shown before FIVE YEARS after date of delivery to the original purchaser that it was defective in materials or workmanship when delivered.

This Warranty shall apply only within the boundaries of the Continental United States and is in lieu of all other warranties expressed or implied.

INTERNATIONAL HARVESTER COMPANY

Group photo of the Evansville Works employees assembled in front of the building, 1946. This may be the day of opening ceremonies for the new works that began production June, 1946. WHS 24144.

These early and at the time, innovative, uses for the coolers realized that there was a large demand for the farm to use refrigeration. The Refrigeration department was created in November 1944 with Eugene F Schneider appointed as general manager. He had joined the company in 1929 as a salesman, becoming Kankakee district sales manager in 1937, was district salesman of eastern USA in 1943, serving in the refrigeration department until 1947, when he was appointed manager of the tractor division.

Subsequent leaders of the Refrigeration division were:

Mr. Layton, who started as draftsman in McCormick Works in 1936 and replaced Schneider in 1947.

Replacing him was Mark Keeler on May 1, 1950, who started with Harvester in 1929. Prior to becoming the refrigeration manager, he held several manufacturing positions during war, was general superintendent of the company gun plant at St. Paul, starting as works manager and later manager of manufacturing for the division at the gun plant.

Charles Harris, former manager of Engineering Refrigeration, was promoted to general manager on May 15, 1952, succeeding Keeler.

The new department was building these cooling boxes at the West Pullman works and had outgrown the space available in this factory.

Magazine Article from a Harvester magazine showing the new 4 cubic foot freezer on the Evansville assembly line. 11 cubic foot model assembly line is in the background. TC.

Evansville

When Harvester set out to find a new location for the refrigeration plant, the criteria was simple, availability of skilled labor, established shipping routes including roads as well as a strong American know-how and progressiveness.

These needs were found in Evansville, Indiana, which many other refrigeration companies also called home. The largest concentration of refrigeration manufacturing added International Harvester to the list of makers.

Through the War Assets Administration, Harvester bought one of the city's best factories on May 16, 1946. The Republic Aircraft Corporation had produced 10,000 P-47 Thunderbolts over a two-year span in this building. The factory was built with funds provided by the Reconstruction Finance Corporation early in World War II for Republic Aircraft. Spread over seventy-one acres, more than half was a concrete surface. Buildings and floor space covered twenty-one acres or 934,000 square feet. The factory buildings were built with specialty glass containing cobalt, giving the windows a light green hue, which assisted in reducing glare. Properties also allowed the glass a level of ultraviolet protection, keeping the heat out and an even distribution of daylight throughout the building.

The flooring was mostly creosote-treated wood blocks, offering non-slip characteristics and reducing fatigue on the employees' feet. Individual processes where extra ventilation was needed happened in booths or canopies, with dedicated ventilation systems to prevent excessive number of fumes to circulate in the rest of the building.

View of the North section of Evansville Works conveyor line during testing of the conveyor. WHS 24495.

The main building had the ability to operate four complete and independent assembly lines. In 1947, one line was for milk coolers, another for 11-cubic-foot freezers and two lines were being built for upright or vertical units. The two lines were built with flexibility in mind, converting quickly from various models. A few years later, the plant would produce cotton picker drums and M1 Garand rifles.

The layout of the refrigeration factory was well thought out, allowing for uninterrupted flow from raw material at the south end to finished product at the north end. The final units were moved on a conveyor belt to a warehouse nearby. Full production of the plant allowed for thirty-two carloads of refrigerators, freezers, and milk coolers every day.

Five hundred employees from the Evansville area were involved in the reconversion of the plant, some of which were trained in the process of the installation and production of the project. These first 500 people employed by Harvester were joined by three thousand more, bringing the total employees to 3,500 at full production.

Production of Units

The first milk cooler at the new refrigeration plant was built June 12, 1946, 27 days after formal transfer of title was made. This speedy changeover was necessary to continue producing thousands of coolers, moving from the West Pullman Works in Illinois. West Pullman, a suburb of Chicago, was experiencing rationed power, preventing the consistent production of milk coolers. The farm tractor division and the refrigeration division at West Pullman collaborated to transfer the essential equipment and material to Evansville to begin production. Ingenuity, improvisation, and interest were the three driving forces to produce the milk coolers in that short of time.

Front of sales brochure for the 1949 year of freezers. TC.

Production of the freezers began shortly after the move to Evansville as well. By the end of 1947, two models of freezers were being built, the first 11FC and in mid-1947, the 4FC was released.

Informational articles about the plant included future planning of two upright refrigerators with single or two temperature settings and vertical freezers. (These were introduced around September of 1953)

The size of the property and the success of domestic refrigeration gave Harvester the confidence to imagine every use possible in a variety of classes and sizes.

Development of a line of refrigeration for the trucking industry was also included. Transportation of perishable food over long distances required lower temperatures to keep the food fresh, protect appearance and nutritional value. Adding the transportation line to the existing commercial and newly formed domestic lines gave International Harvester a complete line of refrigeration services.

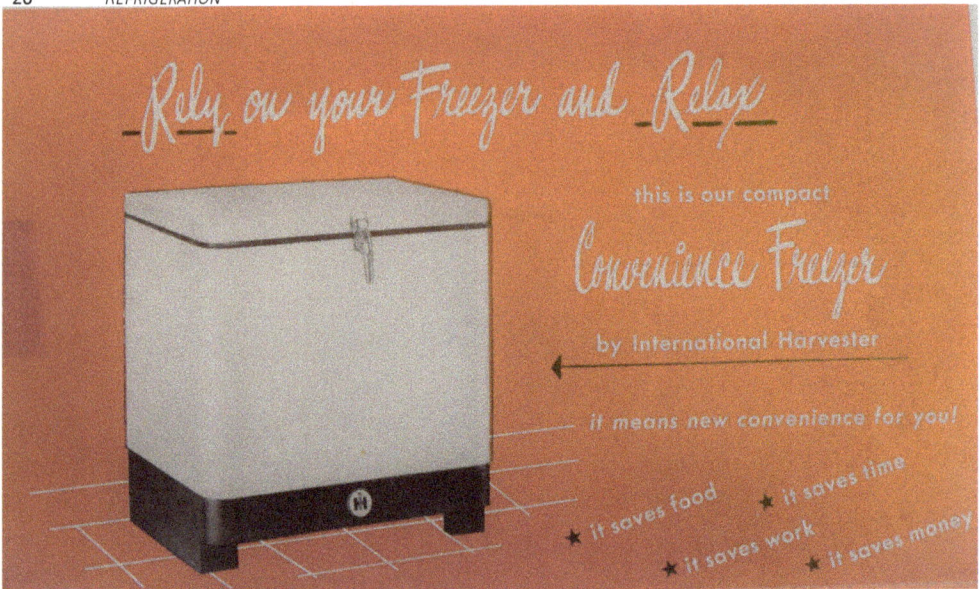

Rely on your Freezer and Relax

this is our compact

Convenience Freezer

by International Harvester

it means new convenience for you!

★ it saves food   ★ it saves time

★ it saves work   ★ it saves money

4FC was introduced as a compact, useful freezer in urban areas. TC.

Harvester felt that offering the four lines of service – household, farm, commercial and transportation – gave them an edge in competition over the other 230 firms already offering refrigeration of some type. Entering the field of domestic refrigeration gave Harvester, for the first time, a product to sell to every home that had electricity.

Even then, Harvester saw the future where we would buy any refrigeration unit ready to plug-in and start using.

The first freezer that was built by International Harvester was the 11-cubic foot Freezer, which was then shipped to Mr. and Mrs. Will Fleming of Ankeny, Iowa through the Des Moines branch in November of 1946.

In 1947, International Harvester introduced a freezer specifically for home use. No longer would the food be kept in the milk house.

A typical kitchen in that era was still viewed as a workspace, a utilitarian environment separate from the dining room. It had evolved from the previous generations of kitchens that first began as little more than a handful of pieces. These usually consisted of a wood-fired cookstove, a cupboard to hold pans and basic supplies like flour or sugar. Centered in the room was a table that doubled as a workstation for preparing anything from fresh vegetables, to cutting meat, to rolling out pastry for pies. There was very little in the form of counter space as modern kitchens today feature. The style of living in 2023 includes the kitchen as the heart, a warm, welcoming environment to gather in and dine casually. Features in these modern kitchens reflect the room as it was almost a hundred years ago. A desk space, now integrated into the counter and cabinets, instead of a separate desk or 'secretary' is one such design feature. Another is a 'breakfast nook' or banquette. Once used as an informal place for eating breakfast, it is now used for every meal. The island that is common in almost every kitchen once was a humble worktable.

Back Page of Refrigeration Brochure showing information about freezers. 1949,
also introduces Irma about this time as a fictional character as spokesperson. TC.

The number 4, for 4 cubic feet, and the number 11 freezer was showcased at the national boys and girls 4-H Congress. Youth from all over the country would gather in Chicago to celebrate their awards and compete for the title of national winner. The attendees were the state winners in different divisions such as sheep, crop science, soil science, sewing, food preservation, farm management and many others. International Harvester was one of the many companies that hosted breakfasts, luncheons, and dinners during the Congress. Harvester was known for holding the outgoing luncheon at the corporate meeting rooms, with a group photo taken at the conclusion of the event. Sadly, the National Congress stopped hosting a few years before I was personally able to attend the Chicago trip. It was often mentioned that it was three days of rural farm kids taking over the city! Many of the major business-es such as Goodyear, International Harvester and other corporations who had their headquarters in the Chicago area, were very generous in hosting this event. This was a great way for them to support and give back to the youth to help create a better world.

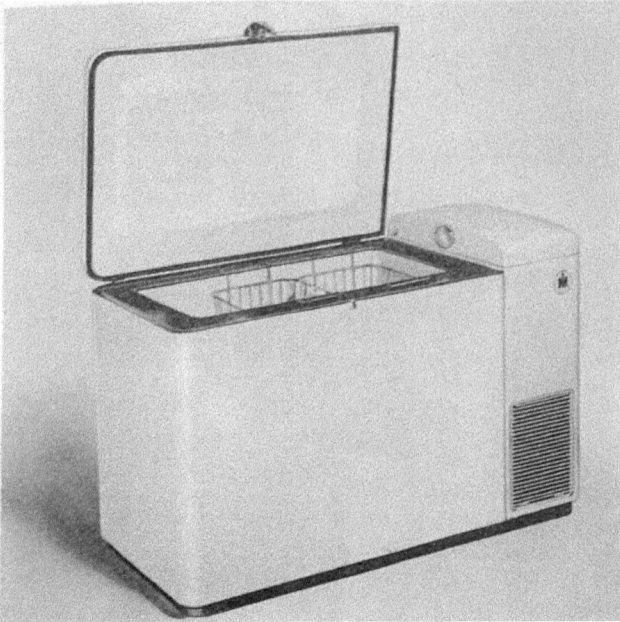

Front View of the 11 FC,
First freezer built by
International Harvester. TC.

Freezers and Refrigerator Production
11FC
This freezer posed a styling statement when it first rolled off the Evansville plant. With smooth simple lines, the appeal was in the modern look. A welded all steel cabinet with rounded corners features a lack of cracks or crevices, keeping the unit easy and quick to clean. A white, lustrous color, the baked enamel surface of the freezer prevented chipping, cracking, or peeling, and gave the cabinet a corrosion -resistant finish.

Starting with serial number 9944C7, two wire baskets and two wire separators, one short and one long, helped keep food organized.

A clip-on thermometer was added with serial number 11200C7.

Units 501-503 were hand built and production started with 504C6.

A different style of variable temperature control was installed on 78 freezers be-tween serial numbers 27352 and 28133.

Serial numbers can be found on the 11FC on the main cabinet inside of the housing unit. 504 to 66149 is located inside on the chimney baffle. 66150 and up is on the main cabinet inside the housing unit.

Insulation between the inner and outer walls, both steel, is made of fiberglass and hermetically sealed to protect it from moisture and air. The springy fibers made from molten glass are light, will not burn, decay, or settle. Operated by a silent and efficient condensing unit, mounted on spring rubber mountings to absorb vibration, the freezer was designed to feature on every farm and kitchen. Fiberglass insulation for the lid, or door, had to be special ordered.

The rubber breaker strip on the top of the cabinet is smooth, and free from screws. Easy to clean, the rubber gasket on the lid closes against it, effectively sealing in the cold. 37 ½ " tall, 58" wide and 29" deep, the 411 pound freezer was placed the in the Fleming family's farm fruit cellar, where it was filled with pies. The smooth working latch assured the operator that the lid was sealed fully.

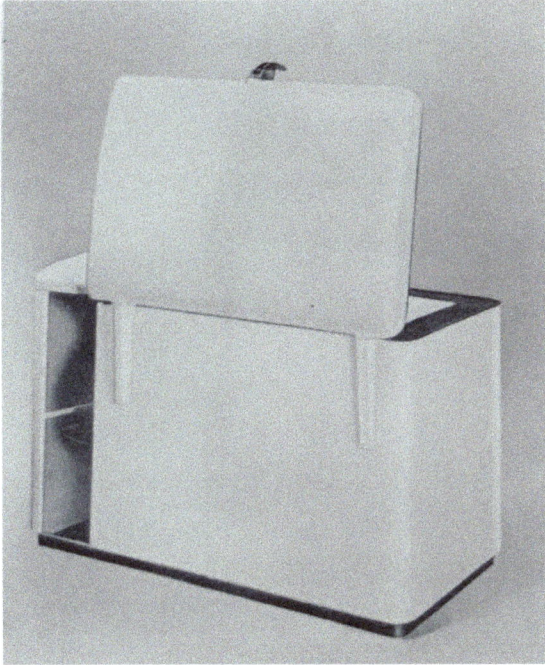

Front and rear views of 11 FC, introduced in November 1946, built until 1950.

View of the 4 FC, built from 1947 through 1949.

The lid operates on spring-counter-balanced hinges, giving it a feeling of floating when opened or closed, preventing the lid from snapping closed quickly. This was the 1947 version of self-closing hinges found in almost every home cupboard and drawer today.

4FC

The 4FC freezer was created to respond to a need for a smaller-sized unit, one that could be in the kitchen, not in a cellar or back porch, as the larger freezers were simply too big.

This smaller size, better suited to city homes and urban families, was Harvester's first venture into the city, and off the farm completely.

Measuring 36 ¼" tall, 33" wide and 26" deep, and weighing 288 ½ pounds, this small freezer was designed to be installed in the kitchen, with a flat lid to double as a workstation, as the style of kitchens in 1947 did not loan itself to acres of counter space.

This was the first household appliance designed for everyone, for everyday use. The tag for the serial number 501B7 and up can be located to the lower left on the rear of the unit.

The appeal of the small freezer was well received, as over fifteen thousand were built in a three-year timespan. The beginning production goal for these freezers was one thousand a month, with increased quotas following that first month.

Total production numbers for Evansville in 1948, including milk coolers, freezers and six months of refrigerators was 44,859 units.

The Model 70, a freezer with 7 cubic feet of storage, replaced the 4FC in 1950.

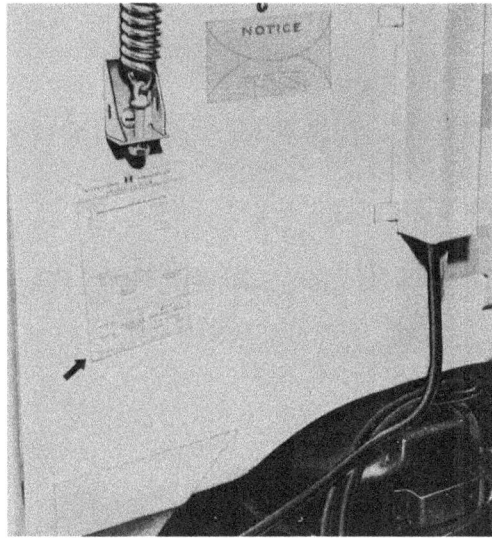

Condenser Serial Number for 4FC located on the frame rail; Serial 501 up.
Cabinet Serial Number for 4FC located rear left of unit; Serial 501 up.

Full Rear view of 4FC, right
Front view of 4FC, below

4FC units were built from November 1947 to 1949, replaced with 70, a seven cubic foot model. The 4FC is the only model International Harvester built in this smaller size.

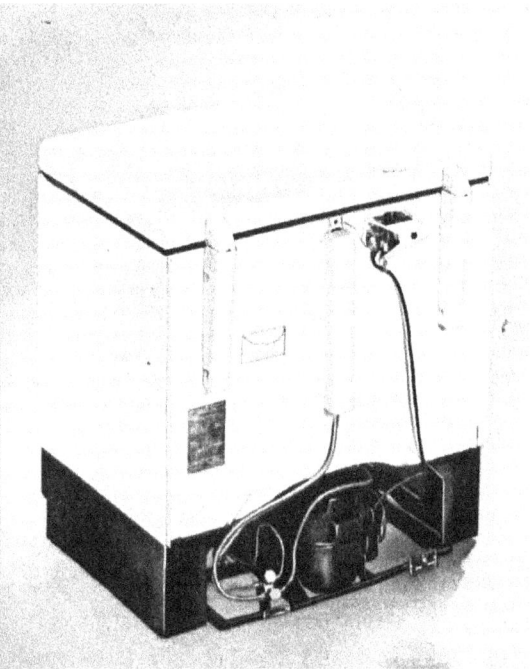

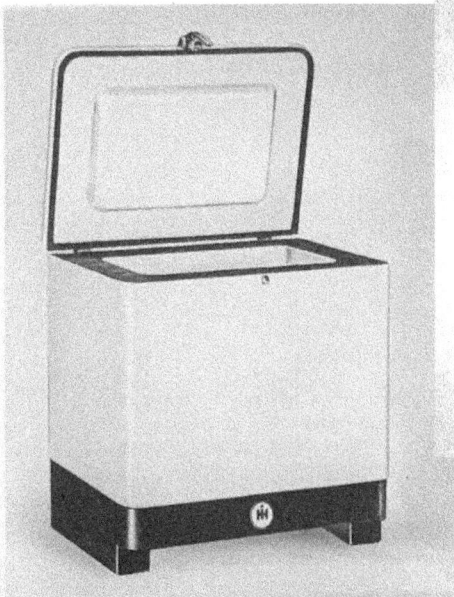

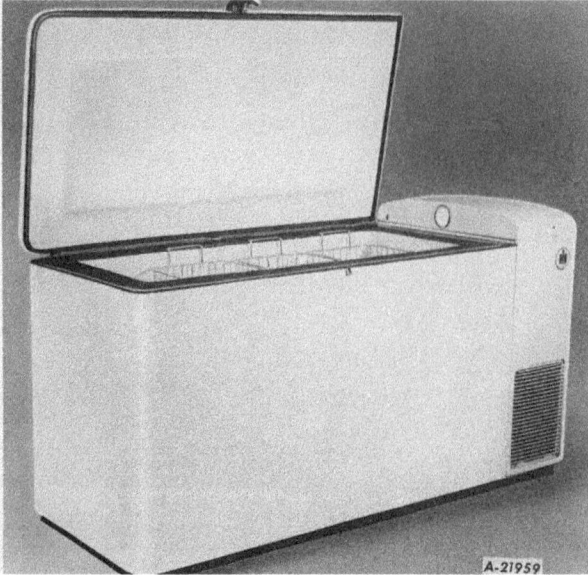

Front and rear views of the 15 FC freezer as shown in the Parts Catalog. Introduced in 1948, and built for two years. TC

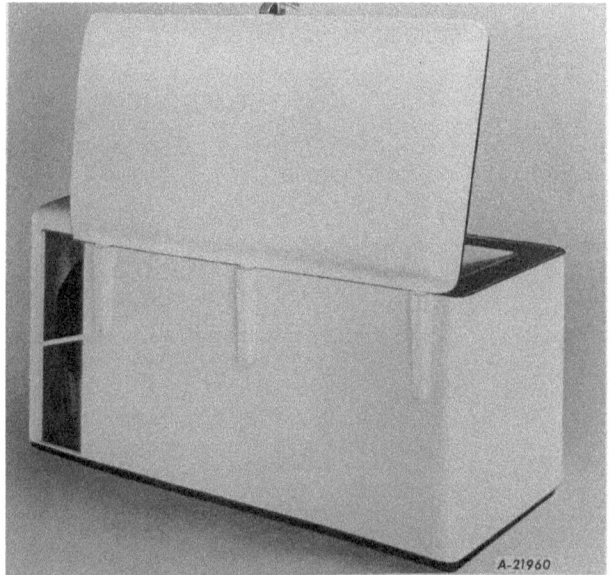

15FC

15FC was added to the lineup for 1948, as a larger version of the 11FC. This unit has all the build features of the smaller two freezers, and was offered until 1950.
15.8 cubic feet arrived for use with three hanging wire baskets, four wire separators, (one long, one short, two medium separators) and a thermometer.

An optional alarm to warn of a power loss could be added if desired.
37 ¼" tall, 73 ½" wide, 29" deep, it could hold 525 pounds of food.

Serial numbers are located on the cabinet behind evaporator cover at top left of unit.

*15FC Cabinet Model
Serial Number - 501 up*

*11FC Cabinet Model
Serial Number - 66150 Up*

Below: Serial Number Location for 11FC Models 504 to 66149

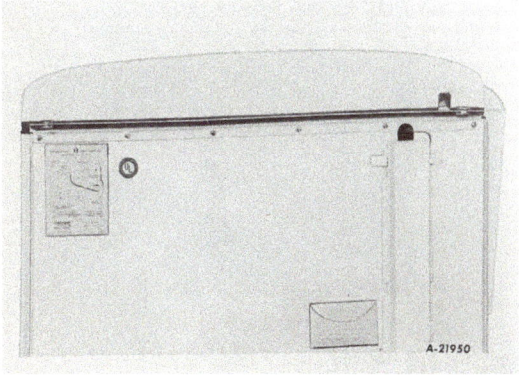

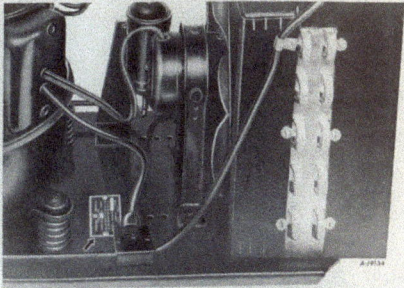

*11FC Condensing Unit Serial Number Locations
Above Right - 501 to 9046
Bottom Right - 9047 Up*

*Left:
15 FC Condenser number located on base of unit. 501 up*

Cover for Refrigerator brochure, 1948. TC.

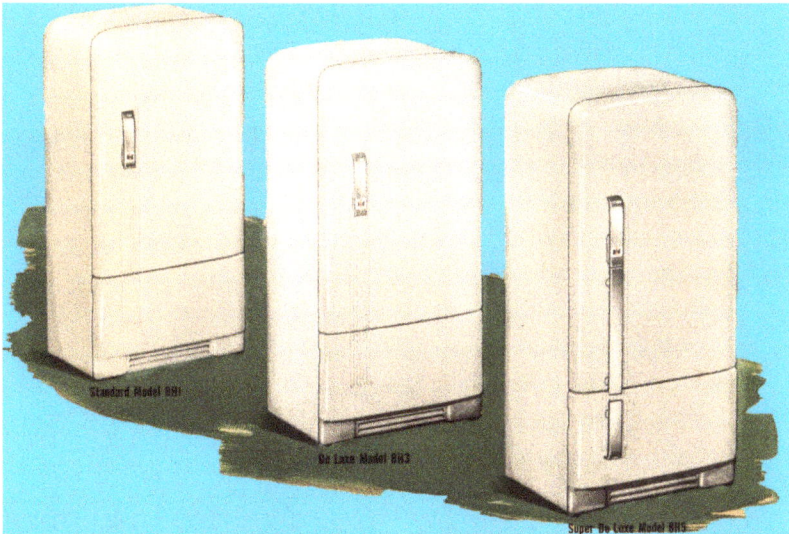

Standard Model 8H1

De Luxe Model 8H3

Super De Luxe Model 8H5

1948 Refrigerators, 8H Series

Following the success of the freezers, International Harvester introduced household refrigerators in 1948. No longer limited to the large commercial coolers, Harvester found a market in helping the farm household become more efficient in their daily activities.

The service bulletin dated December 24, 1947, introduces the three new refrigerators for 1948. The model designates the first letter as cubic feet, the letter H for household, the second digit indicating the style model and the code number separated with a dash, indicating the year. 8H1-48 Standard, 8H3-48 DeLuxe, 8H5-48 Super DeLuxe, all three designed to maintain a regular temperature of 38°F.

All three fridges are 8 cubic feet capacity steel cabinets, in white. The handles on all are die cast zinc and chrome plated with white porcelain interiors. The models vary by the accessories and shelf options.

The evaporator contains a freezer space, with just over one cubic foot of freezer and ice cube tray space. The frost indicator, on the left side of the evaporator shows the maximum amount of build up before the unit is defrosted.

Just months after production began, on the upper left-hand corner, the words "Frozen Storage" were added to the evaporator door. This door began with production of the 8H3-48 Deluxe and 8H1-48 Standard model, the only model to have this door change mid-production was the 8H5-48 Super Deluxe, starting with serial number 7327-B8.

Another feature found on all three units is the breaker strip, silver rose in color, which is easily removed to access the refrigerant lines. These lines are in the insulation compartment directly behind the hammered finish of the Panelyte breaker strip. Updates in July of 1948 to improve appearance and reduce costs for all three models started with the Deluxe 8H-3 Model, serial no. 2342. The Deluxe model was being built with black front flange on the inner liner, gray breaker strips, gray front bin bumper, gray door gasket and red lettering on the control dial.

This update became effective with the 8H-5, Super Deluxe serial no. 10609 and on the 8H-1 Standard with the start of the 1948 production.

View of 8H1 as sold in 1948. Selling points included the amount of space as compared to an Ice Box, reduced melting ice mess, toe relief, and more storage space than a traditional ice box. TC.

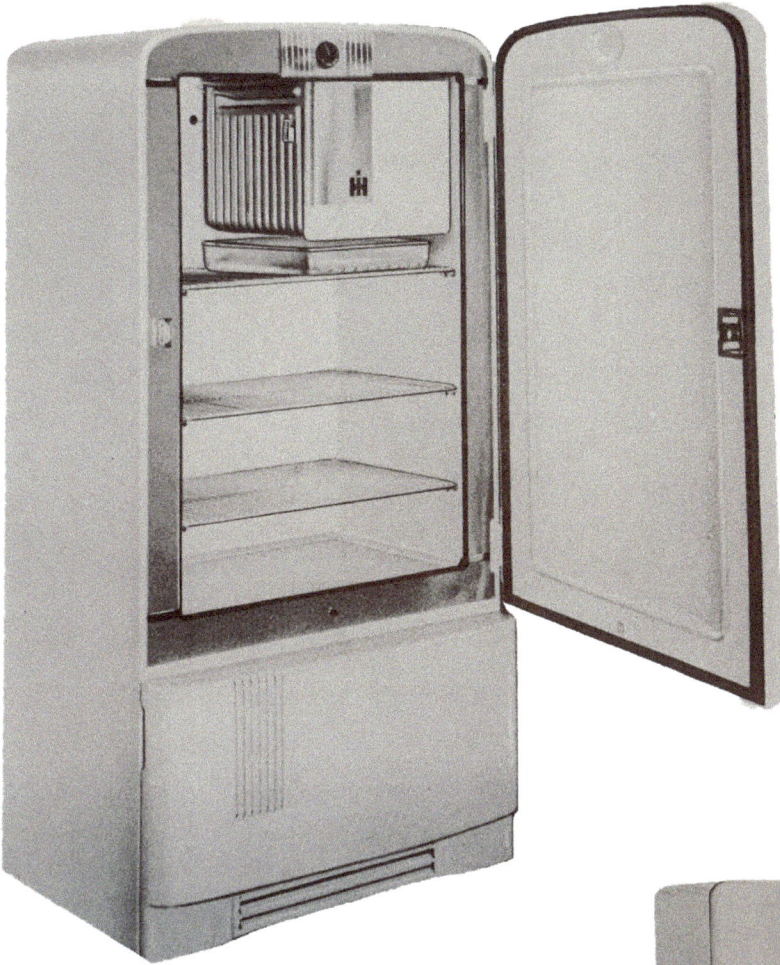

1948 8H1 Refrigerator Serial 501 to 27967.
Features three straight wire shelves and plain freezer door, as well as meat tray, glass.
Built until 1950, and replaced with H- Series and added U-Series line.

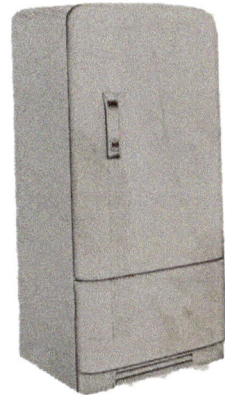

FREEZE FOOD *for better food*...
WRAP IT PROPERLY *for best results*...

Temperature control dial and knob are located in the center of the upper strip panel. The temperature can be set by turning the knob clockwise to lower the temperature. The factory setting are within one or two degrees, and are Defrost 37°F; Vacation, 31°F; Number setting 1 (warmest) 25°F; Number 5 (normal) 19°F; and number 9 (coldest) 10°F. A Ranco temperature control and cover, also saw a updated change early in production, adding a cover over the wire terminals, according to the Underwriter's Laboratories (UL) requirements.

Illuminating from the back panel of the inner liner, a 25 watt appliance bulb in a rubber mounted socket provides light when opening the door, activating the switch found in the lower section of the breaker strip.

Two levelers are standard as well, behind the toe relief area in the front of the unit, to help with uneven flooring and proper door operation.

A balloon type rubber door gasket is used with a positive latching mechanism to eliminate air leaks, retaining the coolness inside the fridge and keeping the heat out. The area inside of the units is packed with fiberglass insulation, as it will not settle, retaining its efficiency in the top, bottom, sides and door of the unit.

Fiberglass insulation, a new concept in 1948, invented accidentally by Dale Kleist in 1932. He was attempting to weld two glass blocks together, trying to create a vacuum-tight seal. The molten glass turned into a stream of fine fibers when a jet of high-pressure air was misdirected. This method was refined in later years and trademarked Fiberglas® in 1938. Insulation of this type was considered a new and modern technology, and not refined as it is today. *(SOURCE: InterNACHI; History of fiberglass insulation by Nick Gromicko and Kenton Shepard.)*

*1948 8H1 Refrigerator
Serial 27968 up. Note freezer door
change and addition of split shelf with
ribbed glass for crisper drawer cover.*

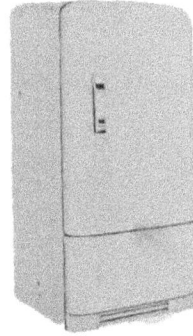

Standard Model 8H1

The Standard Model (8H1) features a completely white finish, three steel wire shelves, white toe relief and silver handle. Formed wire guides on the top shelf hold the 320 cubic inch glass meat tray under the evaporator. Inside the freezer area, four basic, melt out, ice cube trays hold 96 small cubes of ice. The last models, from serial number 27968 and up, replaced the bottom solid wire shelf with a half glass and wire shelf with crisper drawer. Early in production, the wire shelves were also improved, to reduce sharp corners and projections, giving the shelves a smoother appearance, any unit built after 8H-1 serial no. 7800-C8 features these updated shelves.

Model 8H3 refrigerator, DeLuxe Model for the 1948-1949 sale years. Below, View of folding shelf, single for the 8H3 Model, two for the 8H5 Model. TC.

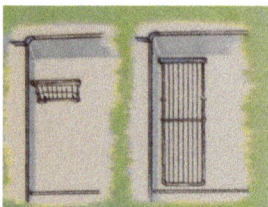

*Model 8H3 refrigerator features half shelf, larg-
er meat drawer, full wire shelf, ribbed glass and
wire shelf as cover for crisper drawer, tip out
storage for non-refrigerated items.*

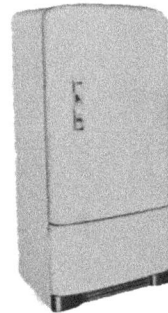

DeLuxe Model 8H3

DeLuxe Model features a black finish in the toe relief space and a pantry bin, for
storage of dry food not requiring refrigeration, beneath the door. The pantry bin
swings out by pulling on the hand recesses at the sides near the top. Wire shelves
are arranged differently than the Standard, for a larger meat try and a crisper pan.
The left half of the lower shelf is glass, serving as a cover for the 855 cubic inch
crisper pan, which slides into the steel supporting channels beneath the glass. The
right half of the lower shelf is wire, the next higher shelf is full width wire. The third
shelf up is left half, half width wire shelf. This is to allow the right side of the shelf
room for a 780 cubic inch white porcelain finish meat tray with glass lid that slides
into supporting channels under the evaporator. Serial number 10893-H8 features
an updated crisper pan and meat tray, for a more refined look. Hinged to back wall,
near the upper left corner of the cabinet is a quarter width shelf that can be folded
back to give extra height for the shelf below to store tall objects. These shelves were
also updated with the start of the 8H3 serial no 6500-B8 units.

Four aluminum tilt out ice cube trays freeze 56 large size cubes, permitting seven
pounds of ice at a time to be frozen.

View of Mother and daughter in front of the 8H5, super deluxe model refrigerator for 1948. left, closeup view of the "Tele-Temp" thermometer, mounted on the inside of the door for the 8H5 model only. TC

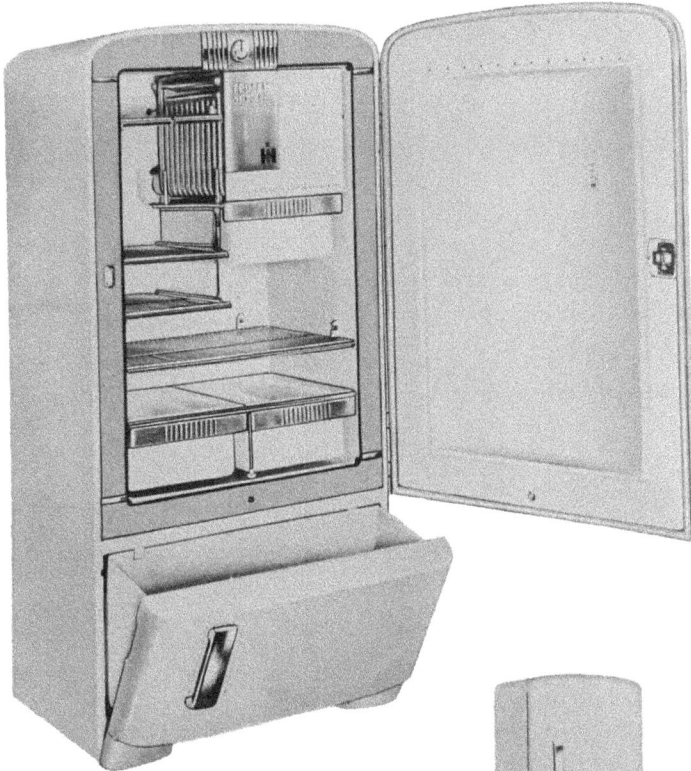

Model 8H5 refrigerator features added flip up shelves, two crisper drawers, extra shelf in the freezer area and tip out storage.

Super DeLuxe Model 8H5

Weighing in at an impressive 306 pounds, the Super DeLuxe Model has all the exterior finishes and features of the DeLuxe Model, with the addition of a chrome plated push bar and pantry bin handle, the same width and finish, matching the door handle and extending it down about two thirds of the pantry bin. The two cubic foot pantry bin is the same size as the previous model. Additional, exclusive features for the Super DeLuxe Model are a second crisper pan, a second glass half shelf to replace the lower shelf half wire, a second half wire shelf above the full wire shelf and a second quarter width folding shelf at the top left half of the box. The glass covered meat tray, evaporator, and ice cube trays are identical in both capacity and finish as the DeLuxe Model. The start of the 8H-5 serial number unit 11827-H8 features smoother shelf radius and overall updated shelf appearance.

A thermometer on the inner door pan shows safe operating temperatures. The Super Deluxe Model features stainless wire with electrolytic polish. Crisper and meat trays are steel with white acid-resistant porcelain enamel finish and stainless steel handles. With the start of 8H-5 serial number 12556-H8 the two trays were updated and redesigned to offer smoother operation.

An extra full width aluminum evaporator shelf can be placed in either of two positions inside the freezer area.

Refrigeration and Electrical System

The hermetically sealed refrigeration system is the condensing unit, evaporator, liquid and suction lines, all manufactured as one sealed unit. The system was tested, crimped and sealed before being installed in the units, preventing any service on the unit after it leaves the factory. The electrical controls can be serviced in the field, and are similar to the freezers in operation, with the electric being provided by simply plugging it into the wall outlet.

With the temperature control points closed, the current goes through the junction block, into the terminals on the overload and starting relay. The current must pass though the overload protector before the common terminal of the compressor, protecting the unit from a surge of power. With the points in closed position, the current is supplied to the terminals and through the starting winding; starting the motor compressor and bringing it up to speed. The current will continue to flow through the windings, operating at full speed until the temperature inside the refrigerator is down to the degree the unit is set at (temperature set-point). The control switch will then open, stopping the motor-compressor, until the temperature control is called to action to repeat the process.

Serial Number locations for these first refrigerators are as shown below.

Model 8H1   Serial No. 501 to 26442;   Model 8H3 Serial No. 501 to 52119
Model 8H5   Serial No. 501 to 46091:   Located at rear of cabinet on
             bottom left corner of control diagram. (see illustration)
Model 8H1   Serial No. 26443 and up;   Model 8H3 Serial No. 52200 up;
Model 8H5   Serial No. 46092 and up   Located at top left rear of cabinet
             on stamped metal plate, riveted to backing. (see illustration)

The condensing Serial number on all 8H models is located on mounting bracket, look at the condenser from the front of the refrigerator. (see illustration)

Location of Cabinet Serial Number
8H1 - Serial No. 501-26442
8H3 - Serial No. 501 to 52199
8H5 - Serial No. 501 to 46091

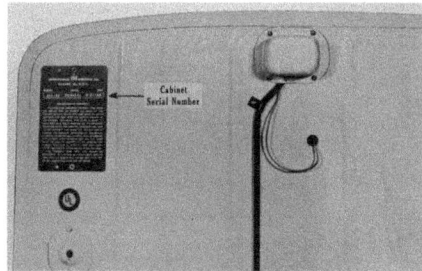

Location of Cabinet Serial Number
8H1 - 26443 up
8H3 - 52200 up
8H5 - 46092 up

Location of Condensing Unit on 8H Models, remove pantry bin to view number located on bracket.

The Freezers introduced in 1950 and built until 1953 were the Models 70, 111, 158

1950 Freezers 70, 111, 158

Model 70
A 7 cubic foot freezer, with a flat top, designed to blend as part of the kitchen. Arriving with one wire separator and one wire basket, this freezer featured a complementing name plate and rectangular handle for the H, HA, U, UA and G-series refrigerators. The serial tag is located at the rear of the unit on the lower left side.

This unit did not come with a lock, it could be ordered special.

Measuring 44 ¼" wide, 36" tall and 27 ¼" deep, holding 245 pounds of food. Built until 1953.

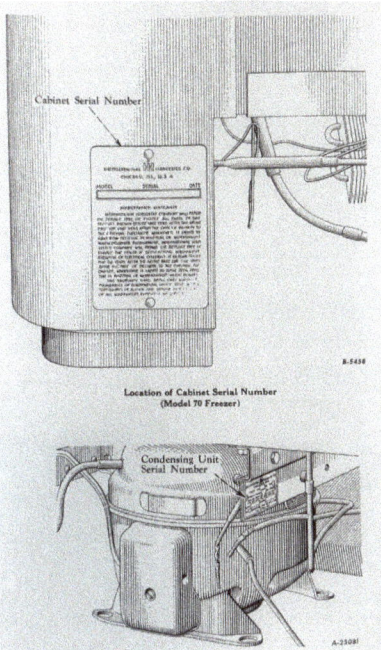

Cabinet Serial Number

Location of Cabinet Serial Number
(Model 70 Freezer)

Condensing Unit
Serial Number

Above: Model 70
Serial number and Warranty tag are found on the bottom right of the Model 70 Freezer, on the rear of cabinet. the Condensing Unit serial tag is on the bracket beside the unit.

Right: Closed view of Model 70 Freezer.

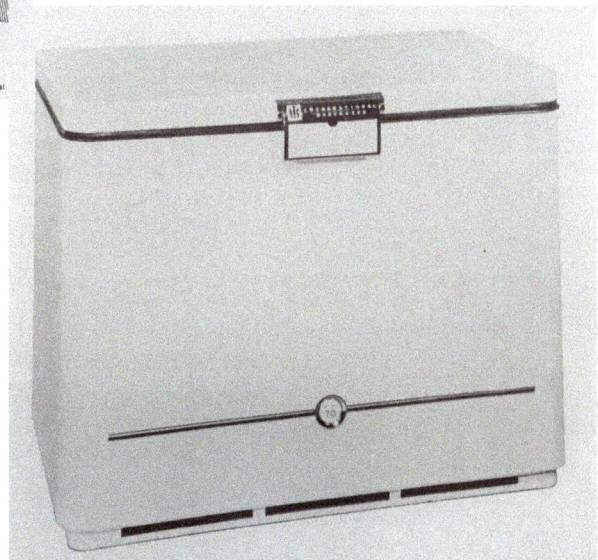

THE FREEZER WITH
- Over-All, Sub-Zero Freezir
  sq. ft. Fast Freeze Area
- Dri-Wall Cabinet
- True Hermetically Sealed

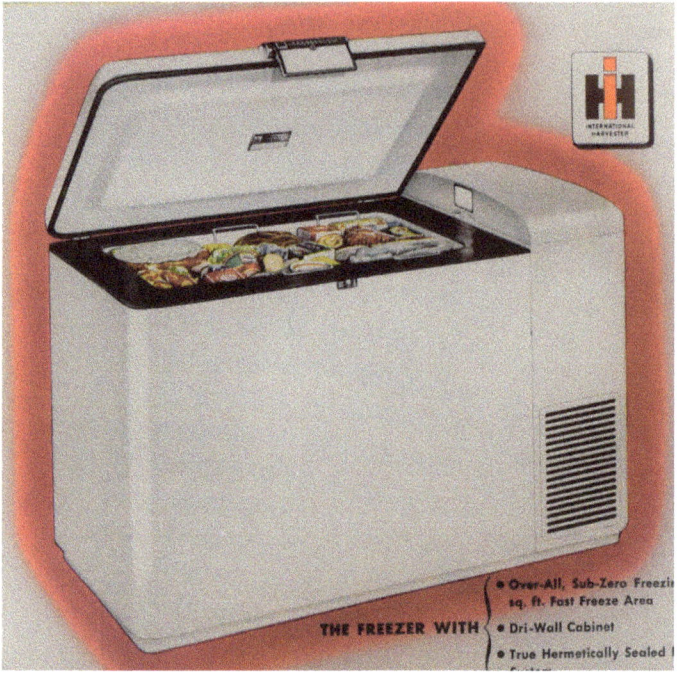

Model 111 Freezer

Feature added is a light that turns on when opened, this 11.1 cubic foot of freezing space could hold about 389 pounds of food.

58" wide, 37 ¼" tall and 29 ¼" deep, it arrived with two wire baskets and two dividers.

The name plate on the inside covered the temperature bulb, and the exterior top showed the inside temperature.

The lockable handle came standard on these larger freezers.

Serial tag is found on the cabinet left side of the condenser unit area.

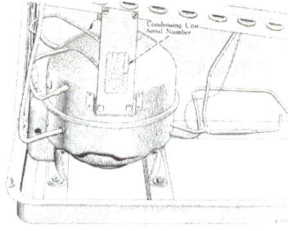

Left: Condensing Unit Serial Number is located on the top bracket of the unit.

Right: view of the exterior temperature gauge for Model 111. This gauge was also standard on Models 158 and 200.

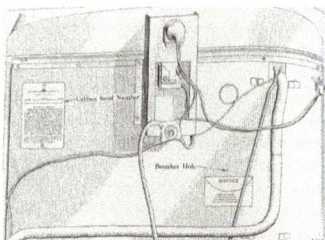

Serial number for the 111 cabinet is located on the left top of the cabinet, found after removing the condenser cover.

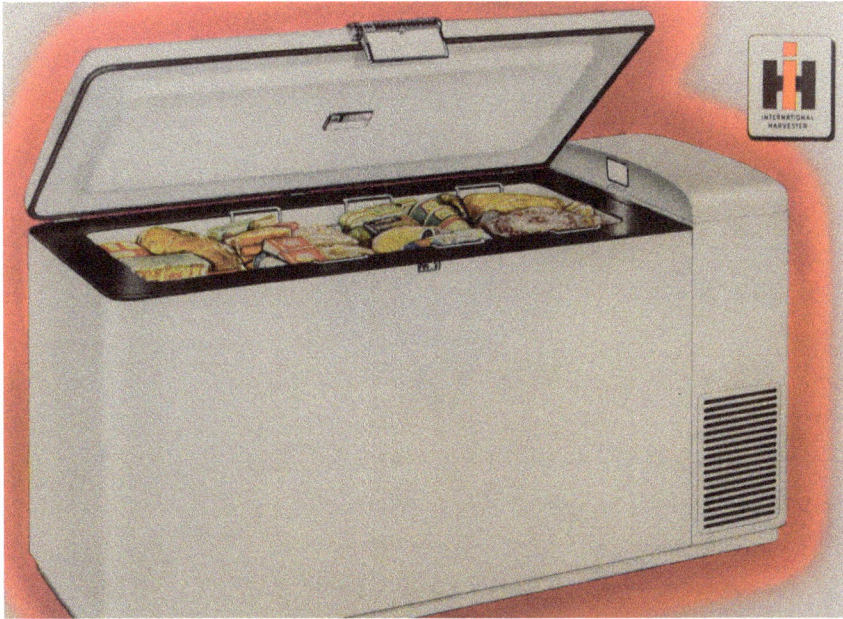

Model 158 Freezer

Holding about 553 pounds of food, this 15.8 cubic freezer came with all the features of the 111 model. Three wire baskets and two dividers, this unit also had an optional extra alarm if power was interrupted, risking food spoilage.

73 ½" wide, 37 ¼" tall, 29 ¼" deep, this freezer was suited in a cellar or covered porch. The exterior temperature gauge provided a convenient way to monitor inside.

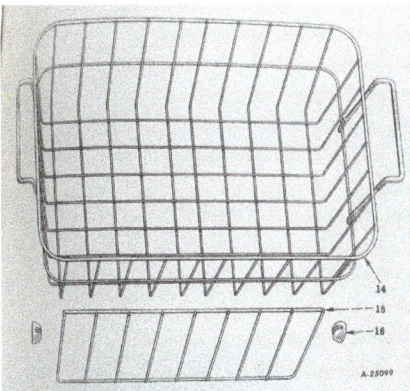

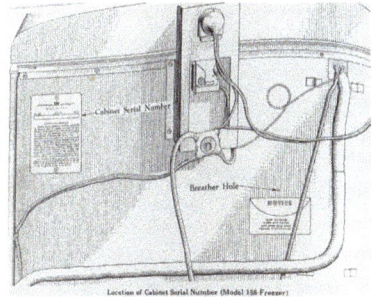

Three baskets, two dividers, an optional alarm to warn of power loss came with 158 Freezer.

Serial and Condenser unit numbers can be found in the same location as 111 and 200.

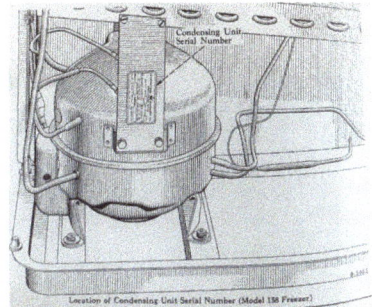

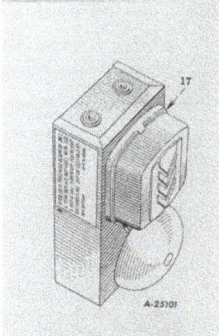

SIX NEW

1950

INTERNATIONAL HARVESTER REFRIGERATORS

Sales Brochure for 1950 Refrigerators, introducing all new styling and options. TC

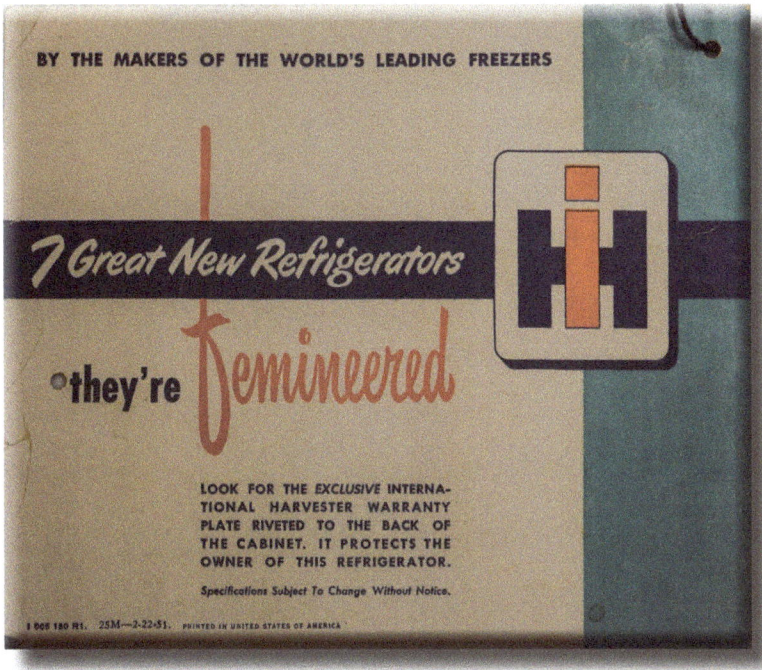

Hang-tag found on sales floors of the refrigerators, December 1950. TC

1950 H Series Refrigerators
H-74; H-84; H-92

The 1950 Refrigerator Models were introduced with a new slogan, "...they're femineered!" The wives had spoken, and the Evansville engineers listened. Mostly. These models featured a full-length door, pantry bin on some models, full width freezer compartment, and the introduction of the bottle opener at the latching mechanism. The compressor is referred to as a "Tight-Wad", a sealed unit that required no maintenance and reliable. Many are still operating in 2023, without any issues.

Building on the previous models, the insulation, exterior finish and construction spared no compromise on the quality. Two coats of white Dulux enamel were baked onto the steel cabinet, with the interior featuring a porcelain – enamel finish, including the acid resistant bottom, for easy cleaning. The temperature control features "Off," "Defrost," "Vacation", and nine cold operation settings. When the unit is in the "Off" position, the interior light will work, while the compressor and condenser will not engage, useful for dealer display.

Serial numbers start with 501, for all models. A welded steel casing permanently protects the mechanism from outside moisture and dirt. Internal spring mounting reduces vibration and noise, and long-lasting operation is reassured with a five year warranty from International Harvester.

Each model has its own features, relating to the unit and cabinet models.

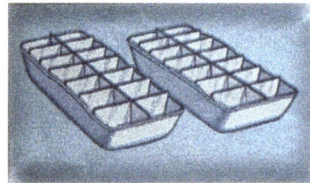

H 74

The smallest refrigerator at 7.4 cubic feet, this was the perfect size for a small family. The Pantry-bin at the bottom has the ability to store 11.2 quarts that do not need refrigeration and easy temperature control with 12 settings.

The freezer compartment can hold 35 pounds of frozen food, 3 pounds of ice and features a self-closing door. This equals to 1,731 cubic inches of space, fully visible through the ice blue transparent door panel. Other features include a meat try that is conveniently located under the freezer compartment, holding 14.5 pounds of meat.

A lightweight see-through ice blue plastic crisper, holding 11.3 quarts of fruits and vegetables, with a split glass and wire shelf. Two more wire shelves give a total of 24 possible arrangements for this unit, allowing for placement of tall bottles and crisper locations.

The door strike features a handy, always ready bottle opener, meaning no more missing openers. A full length, streamlined tapered, ridged door is smooth for easy cleaning, with a self-actuating latch.

The balloon type door gasket offers a tight seal, delivering 1/8 horsepower of unit. Weighing 249 pounds and dimensions of 54 3/8" tall; 25" wide; 28 7/8" deep, this unit average sale price was $214.95. ($2,740.07 in February 2023)

The H 74 Model arrived with a ice blue transparent freezer door, two plastic grid ice cube trays, meat chill tray and a crisper bin in matching ice-blue color. Above, a model shows the pantry -bin for storage of non-refrigerated items.

H 84

The middle size range of the three-unit sizes, the 8.4 cubic foot refrigerator comprises of 17.2 square feet of shelf area.

The full width 'Stowaway' freezer area has room for 4 ice trays, freezing 7.4 pounds of ice and fifty pounds of frozen food.

A self-closing metal front full width freezer door finishes the upper compartment look.

Four lever release ice trays fit inside the drip tray and will remain solidly frozen.

The temperature control panel is integrated with the interior light, which Harvester referred to as a "Diffus-O-Lite."

A damper control for winter or summer operation offers best usage in all seasons.

Two porcelain enamel crispers hold 18.4 quarts of fruits and vegetables. Both pans have ribbed glass covers.

Two and a half shelves offer multiple arrangements, and the meat storage tray is 530 cubic inches, and can fit inside the freezer compartment.

The pantry -bin, holding 26.1 quarts of unrefrigerated foods and beverages, is located at bottom of refrigeration unit.

The 1/8 horsepower motor powers the weighty 307-pound unit, with dimensions of 59 5/8" inch tall; 29 ½" wide; and 29 1/8" deep, including hardware.

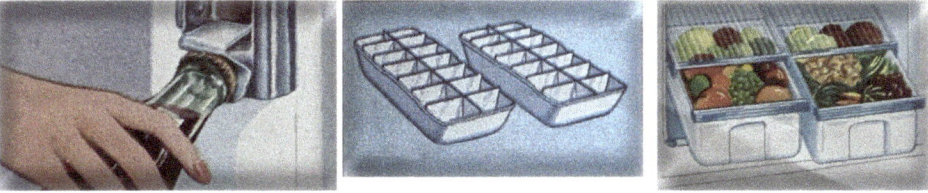

H 84 standard features are two drawer crisper bins, meat tray, four plastic grid ice cube trays, and meat drawer, as well as drawer type pantry for non-refrigerated items.

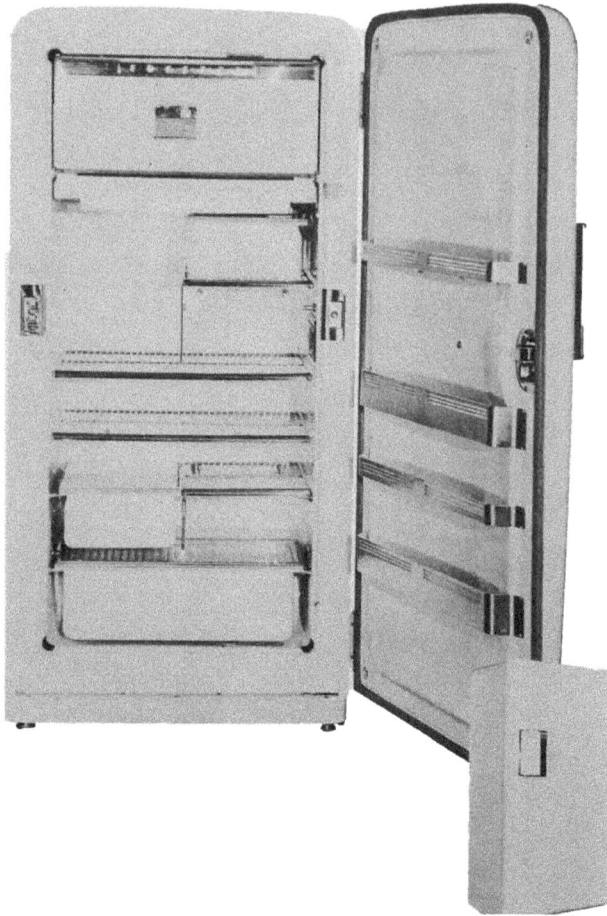

H 92

The largest of the H series, the H92 offers 9.2 cubic feet of storage, four lever release trays, fifty pounds of freezer storage area and a self closing metal door. A full width drip tray under the freezer area and a relocatable crisper pan is where the similarities end with the smaller H84 model.

Additional features are the full width crisper drawer at the bottom of the unit, instead of the unrefrigerated pantry bin. Ribbed glass covers for both bins offer a total storage of 23.3 quarts.

An all new styled feature is the 'Pantry-Dor.' Four narrow shelves attached to the door interior give quick access to frequently used items. The top shelf is dedicated for egg storage, with the other three non-movable trays for Bottle, Dairy and Utility storage, so making a peanut butter and jam sandwich is a breeze for everyone.

In the main compartment, a removable half shelf adds height for extra items, by removing, allows for large watermelons of roasts to fit easily inside. With an overall height of 59 5/8" inch tall; 29 ½" wide; and 29 1/8" deep, including hardware, the outside dimensions are the same as the previous model.

The full length door is easy to clean on the outside, with white enamel baked exterior and a porcelain enamel interior. Under the toe kick are two leveling devices, used when installing so the fridge is level when operating.

H 92 features are two crisper drawers, a meat drawer, chill tray and four lever release ice cube trays. "Diffuse-O-Lite" panel also has temperature control.

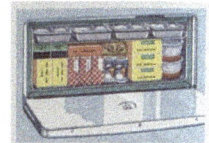

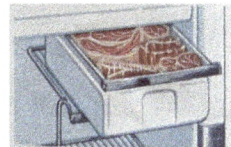

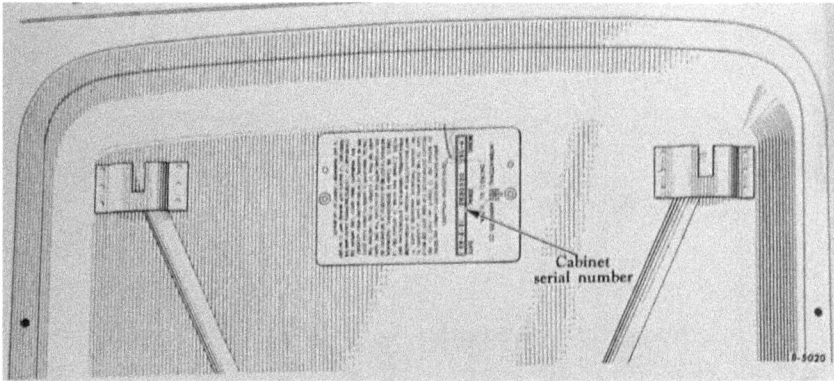

Cabinet
serial number

8-5020

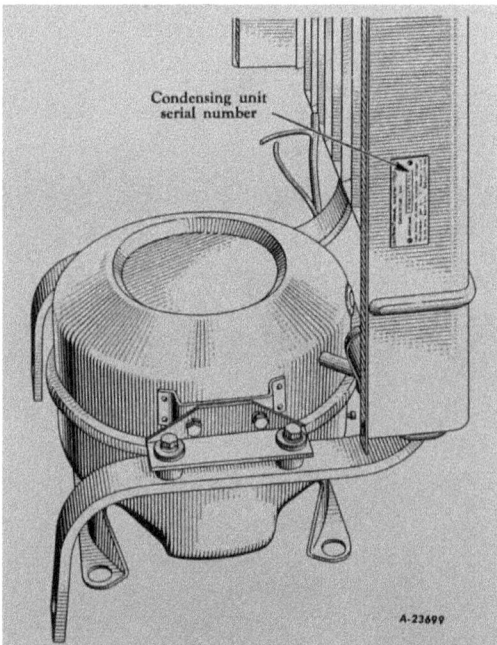

Condensing unit
serial number

A-23699

Serial Number and Condensing Unit Numbers all have the same location on the H and U Series refrigerators.

The Serial Number is on the rear top of the unit, between the wall spacing brackets. This is also the Warranty Tag.

The Condensing Unit Number is located on the side of the housing protecting the condenser coils.

Wall Space brackets held a 3/8" by 3 1/8" Carriage bolt with nut, to keep the refrigerator spaced away from the wall for optimum operation. This wall bracket was used for every model built, including the upright freezers.

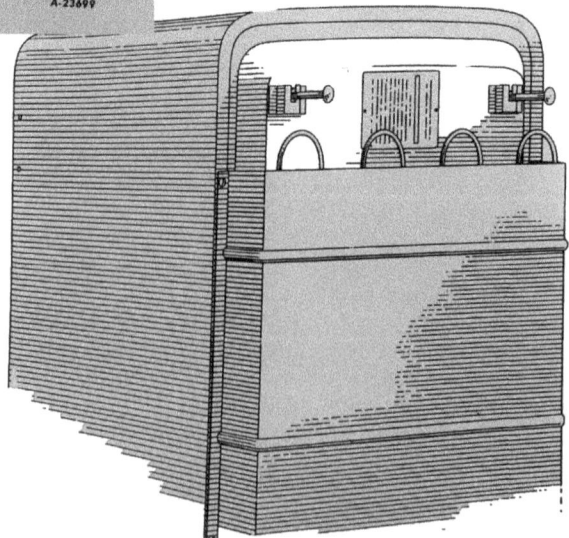

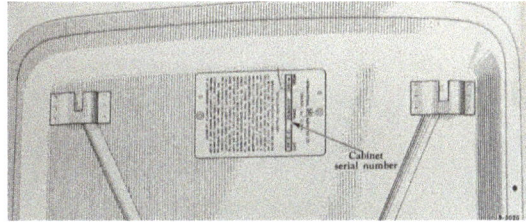

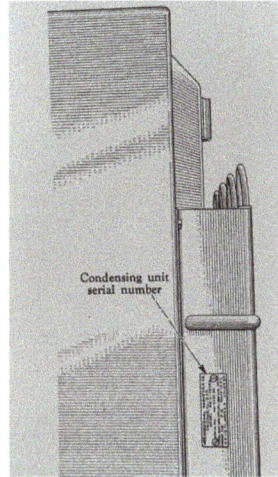

Cabinet serial number

Condensing unit serial number

1950 U Series
U-76; U-87; U-95

The 1950 Refrigerator U- Model was introduced in tandem with the H models. The cabinet and many features were very similar, with the exterior door having a shallow vertical indent starting at the handle and running down the length of the door. The exterior size of the cabinets were the same, with the most noticeable difference in the freezer area. These U-models offered a smaller freezer area, and the unrefrigerated pantry bin was also removed from these models.

These units featured a full-length door, small freezer compartment, and the inclusion of the bottle opener at the latching mechanism remained standard on all units. Building on the previous models, the insulation, exterior finish, and construction stayed the same high standard as the previous models. Irma Harding, a fictional character, started making appearances in advertisements this year as well.

The Serial tag is located at the rear of the unit, between the wall-spacing brackets. Serial numbers start with 501, for all models, as standard for International Harvester. A welded steel casing permanently protects the mechanism from outside moisture and dirt. Internal spring mounting reduces vibration and noise, and long lasting operation from the 'tight-wad' compressor is reassured with a five year warranty from International Harvester.

U 76 Refrigerator for 1950, a entry level refrigerator for small families or singles. TC

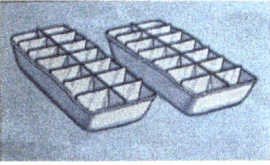

Two plastic grid ice trays, a white plastic meat tray, and a white crisper drawer with ribbed glass lid arrived with every U 76 refrigerator.

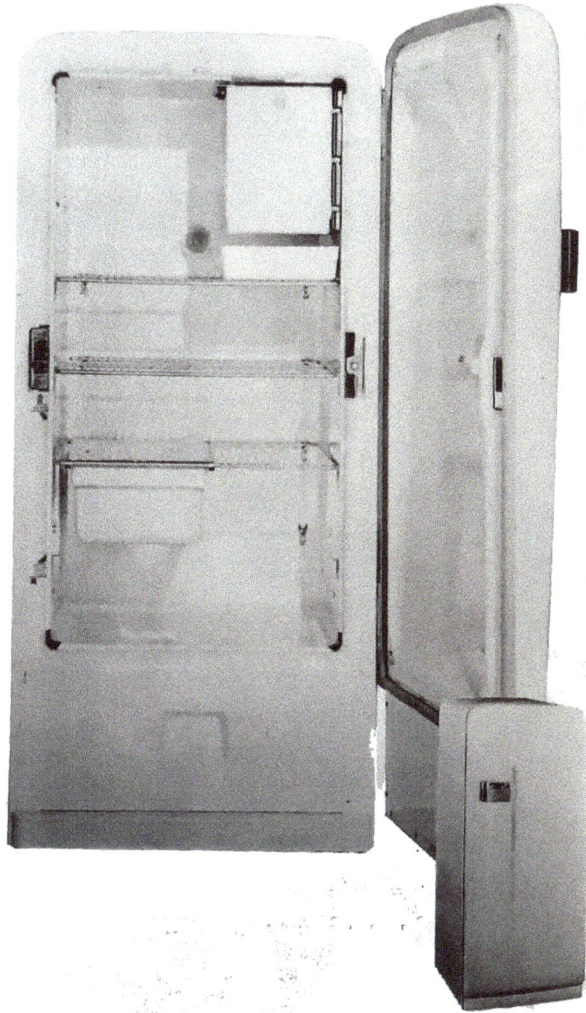

U 76

The smallest refrigerator at 7.6 cubic feet, the half size freezer compartment is perfect for two ice trays, plastic grid type. A plastic defrost tray is free-moving and can be placed anywhere on the two wire shelves or on the third half wire, half glass crisper shelf.

A front panel covers the access to the compressor, behind the full-length door. This model would be considered an efficiency model today, perfect for apartments or homes with one or two people.

With a height of 54 3/8"; 25" wide and 28 7/8" deep, the 249-pound unit from the outside appears to be the sparse version of the H74.

The interior light is located at the upper rear of the cabinet, and the door handle is the same half size. The light did not come with a cover.

Introduced in March of 1950, it was priced to sell at the low cost of $199.95. ($2,548.85 February 2023)

U 87 Refrigerator has an enameled metal crisper, two plastic grid, two lever type ice trays, Egg-O-Mat, diffuse lite with temperature, glass meat tray.
Bottle opener is standard on all models.

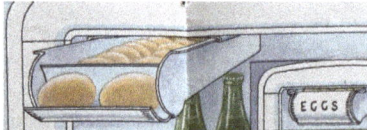

U87 features the shadowline styling on the outside of the door, and repeats the shadow detail in the condenser cover panel on the inside.

U 87

This 59 5/8" tall; 29 1/8" deep; 29 ½" wide 292-pound refrigerator has been keeping drinks cold for over 70 years. Originally sold for $239.95 ($3,058.75 February 2023); the 35-pound, 1,604 cubic inch of freezer space is found on the upper right hand side of the compartment.

Holding 6.6 pounds of ice, using two plastic grid 14 cube trays and two 21 cube ice trays, release cubes easily. 8.7 cubic feet of total food storage area, the U87 features the "Egg-O -Mat," holding 16 eggs, it dispenses one or two eggs at a time, and mounts center of the upper left of the freezer.

A temperature controller is integrated with the light, found below the freezer compartment, the cover is easy to remove and replace the light bulb as needed.

Featuring four shelves offering 14.9 square feet of food storage, comprising of two full wire shelves, one half wire, half crisper shelf and one meat tray shelf. The meat shelf is now full width and formed around the freezer area, so containers can be placed next to the freezer.

The meat tray is glass, with the IH logo in the center. One of only two models that have this style glass meat tray, it is roughly 9" x 13" x 2" in size.

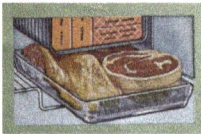

U 95 Refrigerator has glass meat tray, full width crisper drawer, diffuse-o-lite with temperature control, Egg-O-Mat, four ice cube trays and bottle opener integrated in the door latch.

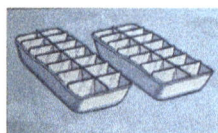

U95

A large deluxe refrigerator with 59 5/8" tall; 29 1/8" deep; 29 ½" wide 292-pounds offers a total 9.5 cubic feet of food storage, and ice freezing up to 6.7 pounds.
A frost depth indicator on the left side of the freezer area lets you know when its time to defrost the unit. Twist the control dial that is integrated with the interior light and defrost quickly and efficiently.
An "Egg-O-Mat" slides out for quick refilling, holding up to 16 eggs at a time. The large freezer compartment keeps up to 35 pounds of ice cream frozen and easy to access. The shelves are conveniently spaced, with the top off-set shelf holding 12 quarts of bottled milk, dropping to a 4" clearance in the meat tray, the second of the only two models to have the IH style glass meat tray.
The total freezer area is the same as the smaller, U87 model, and features the large 14 cube trays, two with plastic grid, two standard grid.
The food shelf storage area of 16.2 square feet is made of three wire shelves, one off-set shelf and one glass crisper cover that can also be used as a shelf. The full-length crisper holds 14.2 quarts of fruit and vegetables. The interior of the door is smooth surface for easy cleaning. The tall rectangular door handle keeps the outside of the unit streamlined with white plastic insert and the IH logo on the top right corner.
Priced at $259.95, the extra features would be equal to $3,313.70 in February 2023.

*Advertising for the color changing door handles, new in 1951
for the UA-87, UA-95, HA-83, HA-84 and HA-92 models.*

HA Series 1951 Models
HA-74; HA-82; HA-83; HA-84; HA-92 Refrigerators

Released in 1951, the HA models offered a new styling option, available on the HA-83, HA-84 and HA-92; interchangeable door handle colors. There were 10 colors to choose from; white; black; light green; dark green; light blue; dark blue; peach; yellow; gray or red.

Serial numbers started at the Harvester standard of 501, the serial tag can be found in the middle rear of each unit. Keeping the styling of the H series, the streamlined shadow style with large lever door style also added the IH shield style emblem to the front of the fridge. Harvester continues with the multi-level price points, so there would be a fridge available for any budget.

These inserts are the beginning of the decorator movement that brought life to the appliances. Kitchens were sparse, consisting little more than a worktable, a sink, the stove and a Hoosier, or a contained workspace unit, for storing kitchen goods. Previously, food was kept in the pantry, as was the ice box. This was usually a smaller room where the cupboards held the serving ware and dishes for the dining room. There was no direct access to the kitchen from the dining room, as previous society customs regulated the hired hands to specific locations. This strict division had ended in most customs; however, the housing layout did not always reflect this in older homes.

The end of World War II brought a housing boom, and the mid-century modern design was born. The kitchen was brought out of the basements and back rooms and into the heart of the home. This was quickly becoming the informal gathering place and the appliances needed to reflect this.

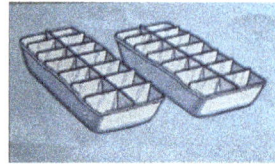

HA-74

The entry level refrigerator to the 1951 line, this seemingly sparse 7.4 cubic foot unit offered a lot of space. Utilizing 13.7 square feet of shelf area, over three zinc plated shelves, a full width chill tray and freezer area, two plastic grid ice trays are included, this was perfect for a small house. Dimensions are 24 7/8" wide; 28 7/8" deep and 54 3/8" tall.

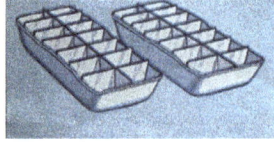

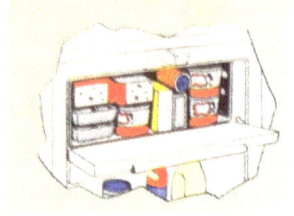

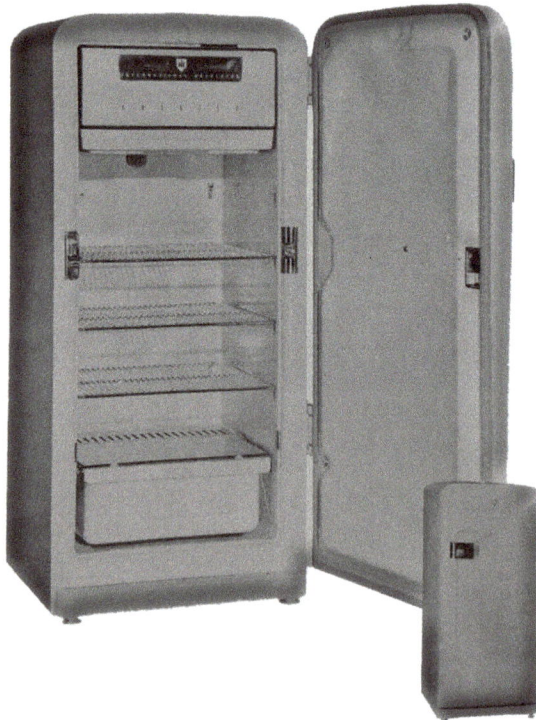

HA-82

Using the same cabinet as HA-74, with the addition of a bottom crisper drawer, this fridge became 8.2 cubic feet, with three chrome plated wire shelves, one glass crisper lid that doubles as a fourth shelf, a full width freezer on top with window view model.

Two plastic grid ice cube trays, chill tray, this cabinet measured 24 7/8" wide, 28 7/8" deep, 54 3/8" tall, suggested retail price was $200. ($2,332.09 February 2023)

HA-82 Arrived with two ice cube trays, a meat chill tray, a full width crisper and a window in the freezer door for 1951.

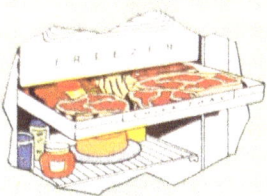

HA-83

One of the models with a choice of color handle inserts, this 8.4 cubic foot refrigerator features 17.2 square feet of shelf space. Three chrome plated shelves, a half-width crisper drawer with ribbed glass cover, and a removable half shelf make this a big value size refrigerator. The full width freezer has a handy flip down door and three ice trays making 42 cubes will stay frozen in the chill tray below the freezer. Dimensions are 29 1/2" wide, 29 1/8" deep and 59 5/8" tall. Weighing 291 pounds, the list price for this model was $220. ($2,565.30 Feb 2023)

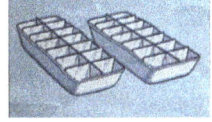

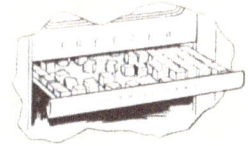

HA-83 featured three plastic grid ice trays, chill tray, crisper drawer, half shelf, and changeable door insert as standard.

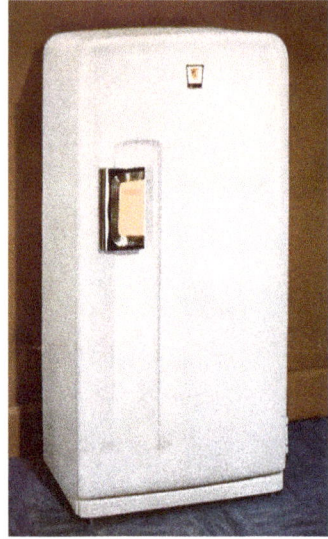

HA-84

Using the same cabinet, color handle option and exterior dimensions as HA-83, the interior on the HA-84 features two crisper drawers with ribbed glass covers, and a dedicated meat drawer fastened below the chill tray. 16.7 square feet of shelf area comes from the three chrome plated shelves, and in-door storage.

Specific egg storage and two door trays give quick access to small items.

The chill tray below the freezer is slightly deeper than the HA-83 and has a capacity of 6.7 pounds of ice, in 60 cubes from two aluminum lever action and two plastic insert ice trays provided.

29 1/2" wide, 29 1/8" deep, 59 5/8" tall, suggested retail is $264, or $3,078.36 in February 2023.

HA-84 Features were two lever action ice trays, two plastic grid ice trays, (four total) two crisper drawers, a meat drawer, deep chill tray and pantry door storage for 1951.

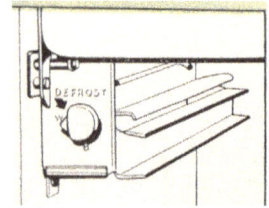

HA-92

The refrigerator with all the features and options for 1951, stainless steel shelving with a safety edge trim, preventing snags on all three full shelves and one removable half shelf creates most of the 18 square feet of storage. An adjustable damper control for efficient summer or winter temperature operation is introduced as a new feature.

The in-door storage features the egg, dairy, bottle and utility trays, adding eight feet of storage. A separate in-door butter keeper featured adjustable temperature control, spring hinge door and one pound glass tray.

Freezer area of fifty pounds of food with a double walled, insulated tray type door and chill tray for 64 ice cubes or 7.4 pounds of ice, from four level style aluminum trays. Crispers were full length, and half width, both with ribbed glass covers, and the last of the changeable door inserts.

This massive refrigerator weighed 302 pounds empty, and measured 29 ½" wide, 29 1/8" deep, and 59 5/8" tall.

Price for this model was $296. ($3,451.50, February 2023)

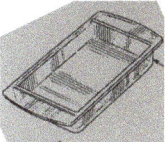

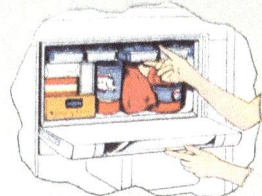

Glass butter dish, one full width crisper, one half crisper, a meat drawer, deep chill tray, pantry door shelves and four lever type ice cube trays were standard for this HA-92 deluxe refrigerator.

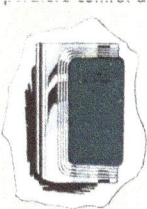

UA-87 featured color changing door insert, meat drawer, crisper, Egg-O-Mat and three plastic grid aluminum ice cube trays.

1951 UA Series
UA-87; UA-95

Harvester continues to offer a refrigerator with a smaller freezer in these two mid-range sized units. Intended to be matched with a deep freezer, these refrigerators continued to be filled with specific options.

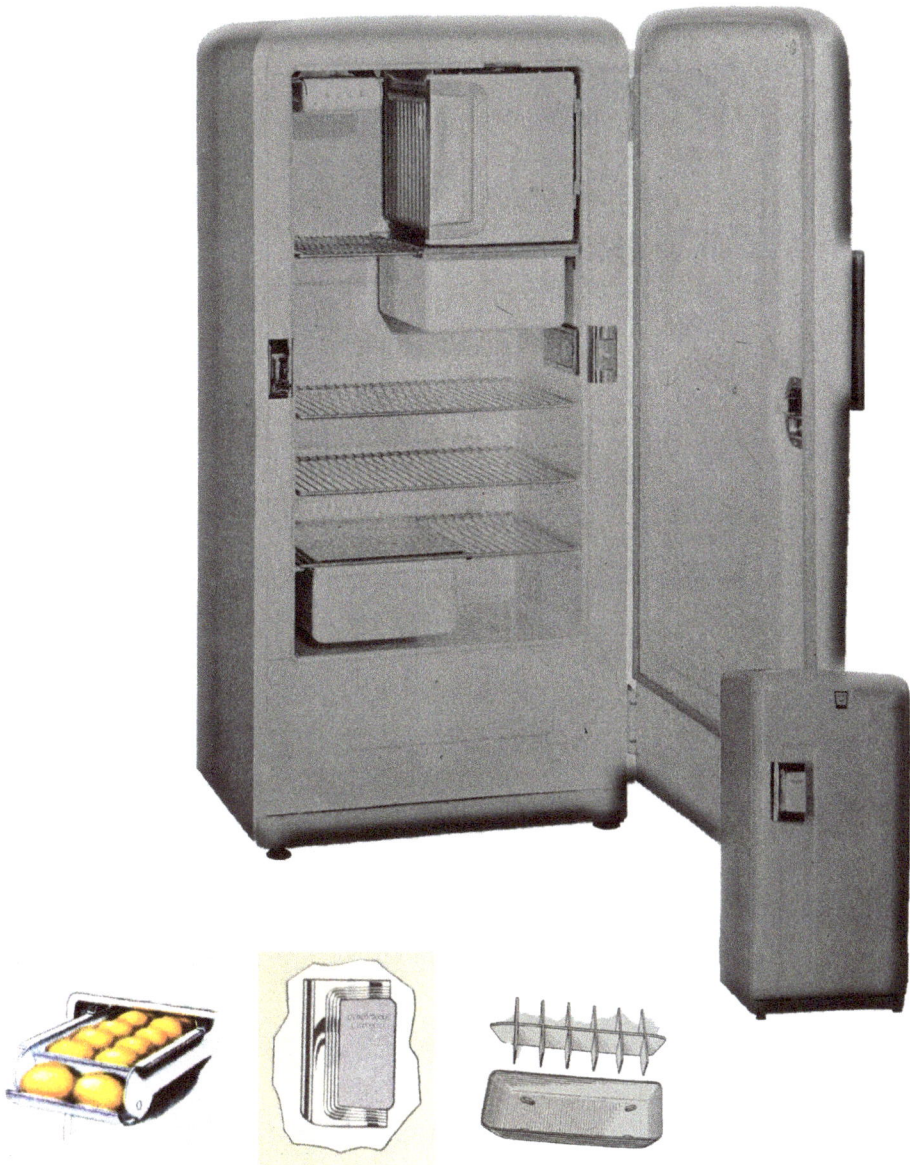

UA-87

This refrigerator introduced in 1951 had very little change from the U-87 model. Updates to the color trim on the crisper drawer, and the meat tray was the larger metal style. The Egg-O-Mat, a 16-egg dispenser was still available as was the color changing door handle insert. Bottle opener remained a standard on all door latches, 1/8 horsepower compressor to keep everything cold.
A suggested cost of $216, ($2,518.66 Feb 2023) made this an affordable option for many homeowners.

UA-95

The 9.5 cubic foot refrigerator is the largest in food volume and meat drawer capacity and smallest in freezer space, with the same 35 pounds of frozen food as UA-87. Third in price range at $240 ($2,798.51 Feb 2023), it suited the homeowner with a large family and a matching deep freezer. Improvements from the U-95 model was a larger meat tray holding 16.8 pounds and three in-door shelves. All other features remained unchanged, chrome plated shelves, four ice cube trays, (two lever and two plastic insert) 14.2 quarts of fruit and vegetable crisper across the bottom.

UA-95 Featured changeable door insert, Panrty door, Bottle opener, full width crisper, meat tray, Egg-O-Mat, and four ice cube trays, (two lever and two plastic grid)

A-29852

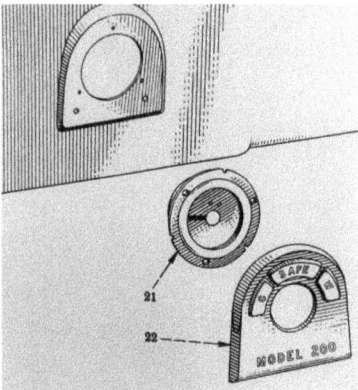

Model 200 Freezer

Introduced in 1952, this large freezer offered a monitoring light, 20 cubic feet storing about 700 pounds of food.

37 ¼" tall, 73 ½" wide, 29 ¼" deep, the lid locked with a key integrated in.

A diffuse-o-lite and temperature control with exterior monitoring of the interior temperature were also standard.

The freezer arrived with four ice cube trays and wire tray to hold them, a covered ice cube container and plastic hostess tray, three wire baskets, and two metal dividers.

Weighting 530 pounds, the 1/3 horsepower motor kept temperatures at the desired setting, ranging from 0° to -10° F. an optional alarm could also be ordered.

SEPARATOR AND BASKET

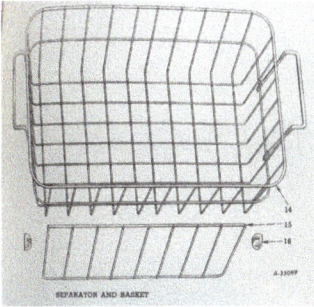

The Hostess Tray and Ice Cube Container was offered for the 1953 L-Series Freezers and the Model 200 Freezer. Offered for the one year, it was a teal blue in color, made especially for International Harvester. The serving tray was designed to fit as a cover for the ice cube tray rack when not in use.

Serial and Condenser Unit Numbers can be located on the cabinet, under the condenser cover.

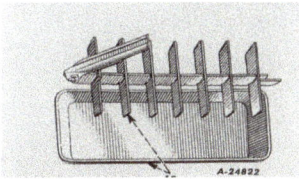

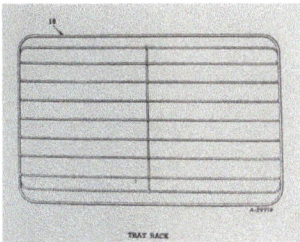

A-24822

TRAY RACK

200

THE FREEZER WITH
- Over-All, Sub-Zero Freezing . . . Over 37 sq. ft. Fast Freeze Area
- Dri-Wall Cabinet
- True Hermetically Sealed Refrigerating System

1952 G-Series

Exterior units featured a virtually unchanged look, with a slight indent or "Shadow-line Styling" feature on the surface of the door. A smaller one piece round emblem with the IH shield insert, and three to five decorative gold buttons with large rectangle handle as trim.

A new feature on the G-85D and G-93D models was the Tri-Matic defrosting feature. This defrost feature could be set three ways, as an automatic option, cycling at 3 AM for 15 minutes, or manually at the hour of choice, when refrigeration and freezer would not be disturbed and lastly, conventionally with the door open and the refrigerator is empty of food for cleaning of the interior walls. A safety feature installed prevented the "diffuse-o-lite" from turning on when the refrigerator was in defrost mode.

For the second year, the decorator handles, now with gold color insert were offered, bringing the total color choices to eleven. The interior color is a soft spring green in shade, reflecting the color trends of the era. Bottle openers in the door latch were continued as a standard item, with the G-85, G-85D, G-93 and G-93D bottle opener models magnetic.

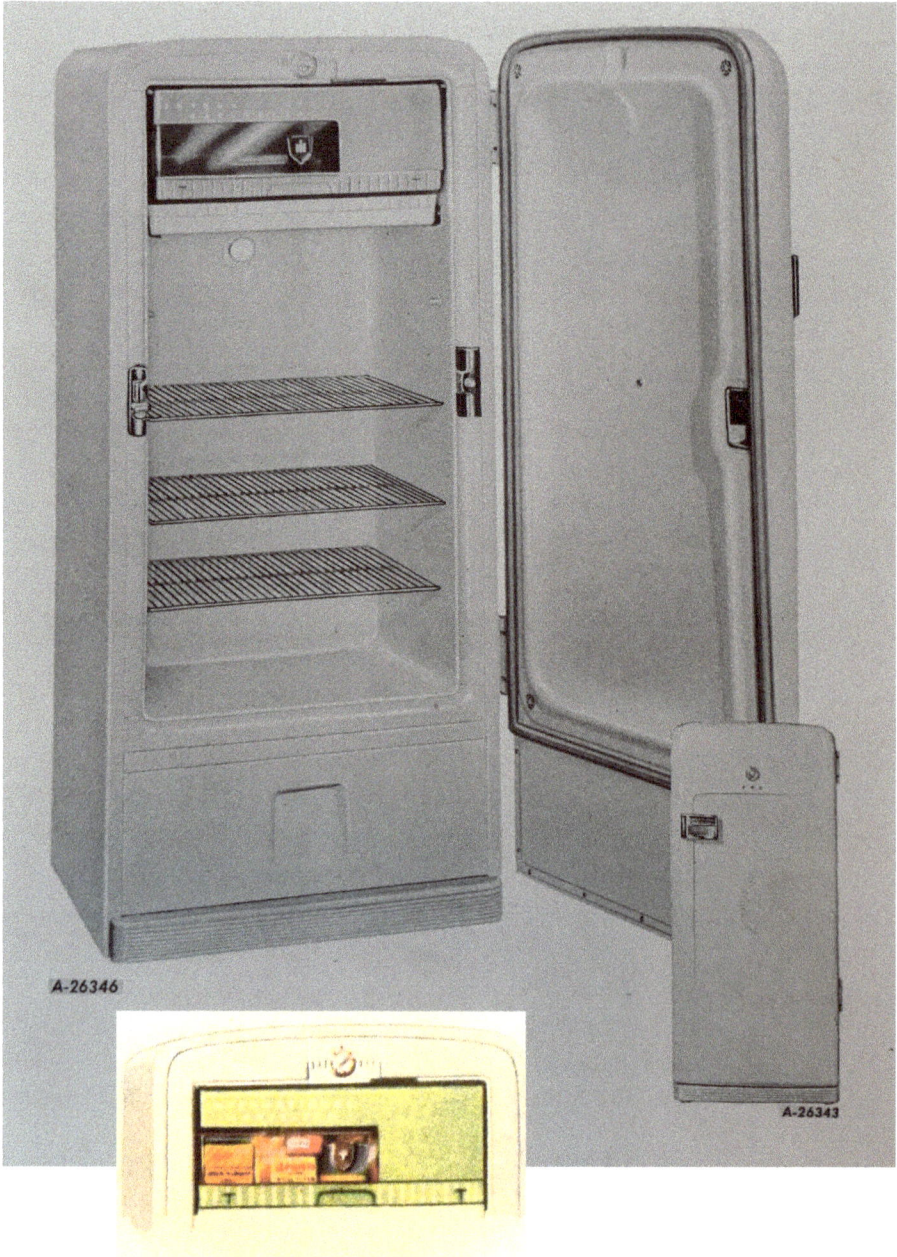

A-26346

A-26343

G-74

The smallest in the G-series, the G-74 featured a full width green door freezer compartment with a partial window to view goods, called the "E-Z-VU Window". This 7.4 cubic foot of food storage area features 13.7 square feet of shelf space, froze 3.4 pounds of ice, held thirty five pounds of frozen food and a green plastic meat tray. The ice cube trays are a plastic grid, releasing with a flex of the grid.
Interior remains white, with three zinc plated adjustable shelves.
Dimensions are 24 7/8" wide, 28 7/8" deep, 54 3/8" high.

G-74 arrived with Spring fresh green freezer door and matching chill tray, and two plastic grid ice trays.

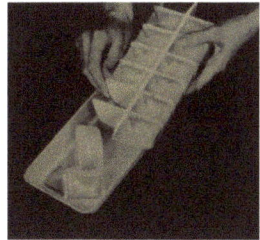

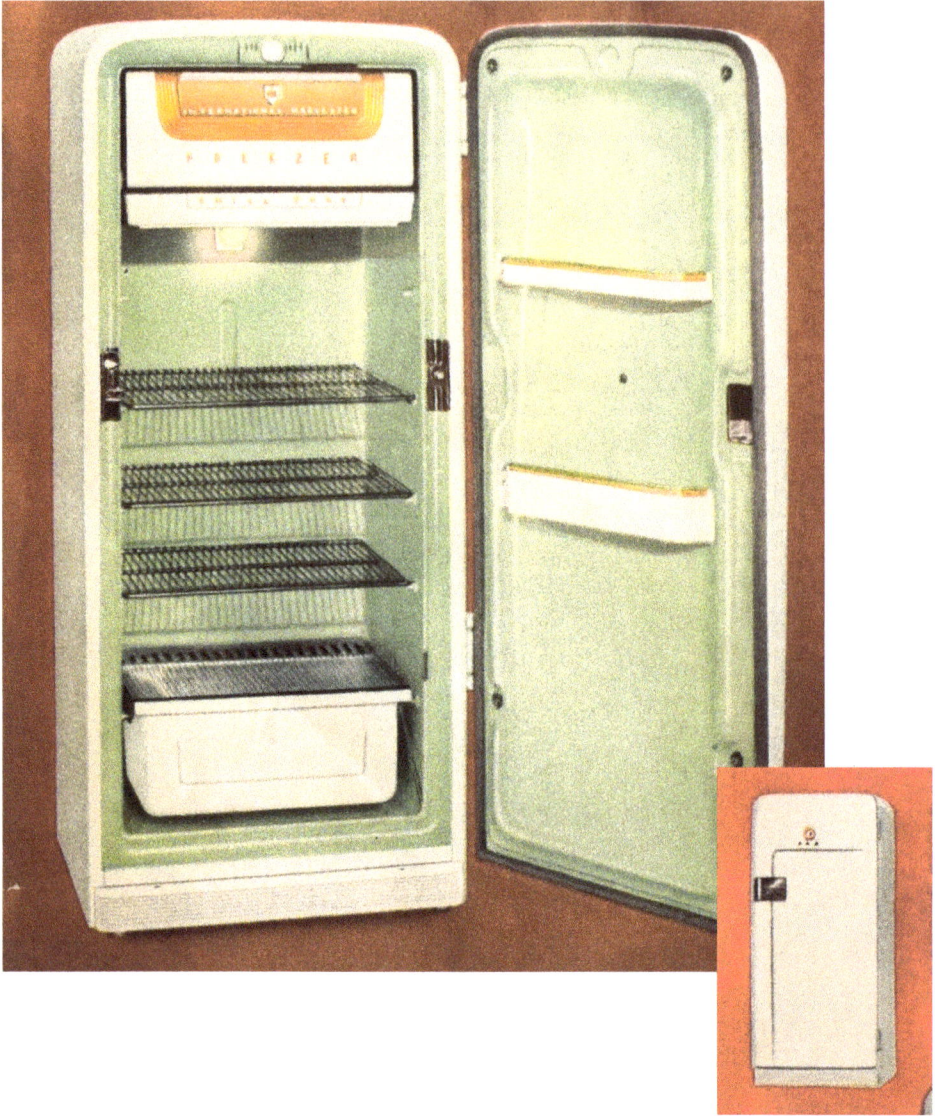

G-82

Full color interior with gold trim accents lends itself class to this 8.2 cu foot refrigerator. This second of the two small units, Harvester added a level of luxury to the one-person sized refrigerator.

The door panel has two trays, one for eggs and another for bottles. Even the breaker strip is green for a continuous look, highlighting the white freezer door, full width crisper and door trays. The egg tray holds eleven eggs in the slots and stacked up to sixteen. The three full width shelves were chrome plated, with an additional ribbed glass shelf covering the crisper.

Dimensions are 24 7/8" wide, 28 7/8" deep, 54 3/8" high.

G-82 features include pantry door, two plastic grid ice cube trays, chill tray and full width crisper.

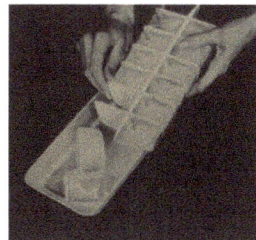

Model G-82 Household Refrigerator

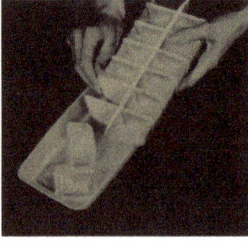

Three plastic grid ice cubes, chill tray and crisper drawer arrived with this G-84.

G-84

Lowest priced in the large class of the series, with the available 'color-key' door insert. Weighing a total of 301 pounds, this unit has a white interior, features the large full width freezer and white meat chill tray. The self-closing freezer door doubles as a shelf when rearranging or loading the freezer compartment. Three large cube plastic grid trays provided 42 ice cubes, for 5.1 pounds of ice. Two shelves in the pantry door, one for eggs and a lower one for bottles and frequently used jars. Three chrome plated steel full width and one half width removable shelves form part of the 17.5 square feet of shelf area. A porcelain enamel crisper with glass lid holds fourteen and a half quarts, leaving room to the side for more shelf space. Dimensions are 24 7/8" wide, 28 7/8" deep, 54 3/8" high.

The addition of Gold to the color changing handles for the 1952 lineup gave the refrigerator eleven decorator shades. Trimatic defrost, and spring fresh green interior updated the look.

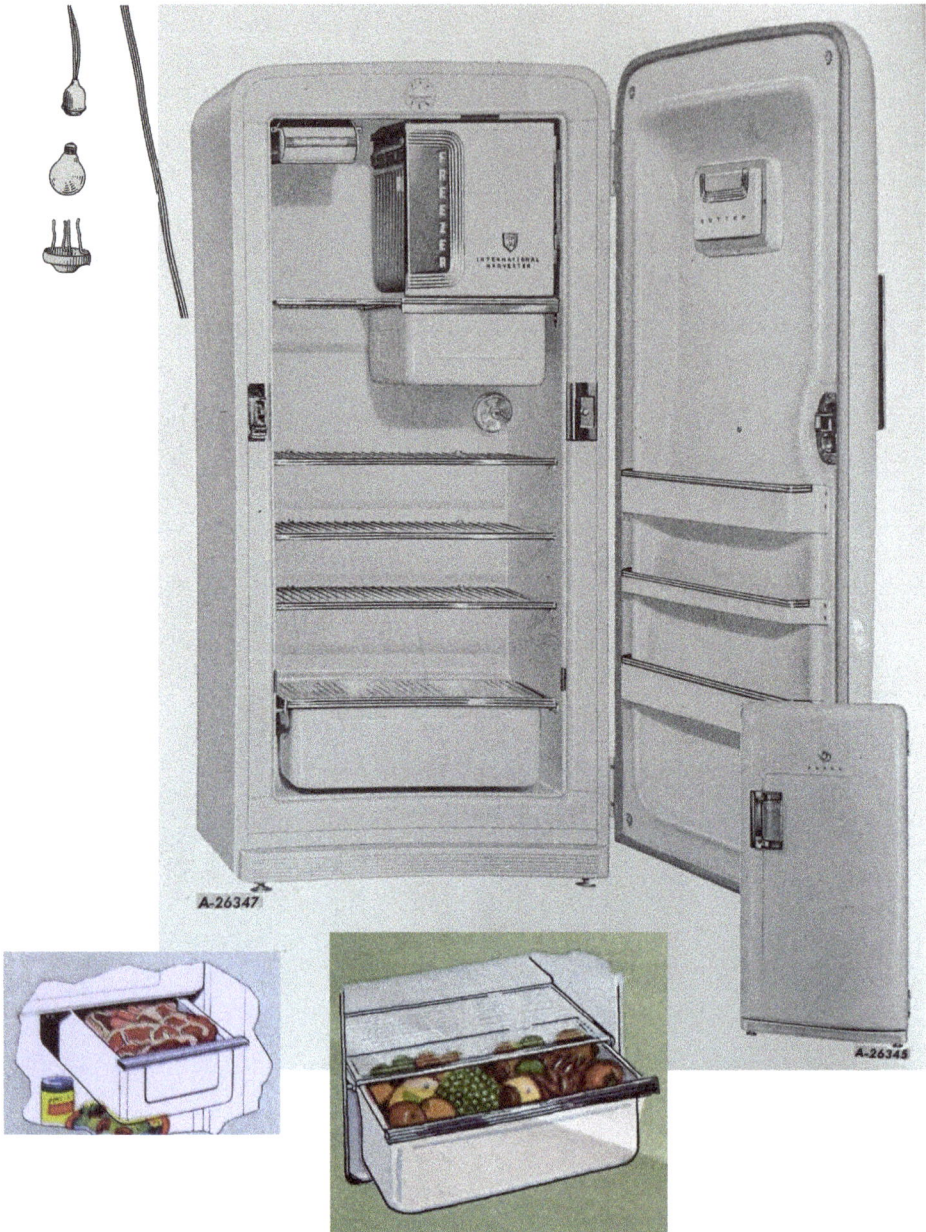

A-26347

A-26345

G-95

The only refrigerator in the series with a half size freezer compartment, this 9.5 cubic foot capacity is the largest in the series for 1952. This refrigerator was designed to be a companion to the deep freezer for families that used perishable food in large quantities. With an all white interior and four chrome plated steel shelves, the trim found on the freezer door, Egg-O-Mat, three pantry door shelves and in-door butter keeper was gold. The freezer held four ice trays, two of the plastic design and two aluminum lever release, for a total of 56 ice cubes. The butter keeper held a full pound at ready to use temperature with a separate control. Spring assisted and hinged at the bottom, it would not close against the butter.

G-95 arrived with four ice cube trays (two plastic release and two lever release), full crisper drawer, meat drawer, diffuse-o-lite, Egg-O-Mat, and one pound butter tray.

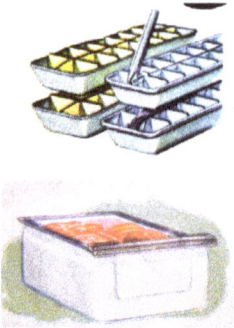

G-85

This all-white interior has many features of the previous models, and more. Crammed into 8.5 cubic feet, a large, full-width freezer, one egg tray in the door and two pantry door trays begin the list of conveniences. The insulated, white full width chill tray under the freezer would hold up to 60 cubes, or 7.1 pounds of ice from two plastic and two lever release trays. A 10.8 pound meat drawer was installed below the chill tray. Two chrome plated steel shelves and a glass-ribbed crisper shelf offered 17 square feet. Two crisper drawers and a in-door butter keeper finished out the features.

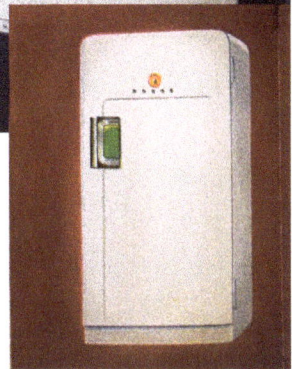

G-85 Features one pound butter tray, two plastic grid, two lever grid ice cube trays, meat drawer, two crisper drawers, chill tray and pantry door shelves.

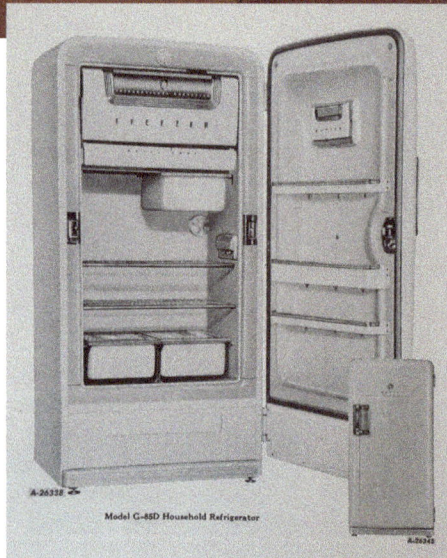

Model G-85D Household Refrigerator

G-85D

The same layout and size as the G-85, this unit featured a green breaker strip, white interior and green door liner. The "Tri-matic" defrost option, four lever action ice trays, cover for the meat tray and a defrost insert in the chill tray complete the up-graded styling changes. Dimensions are listed in rear of book.

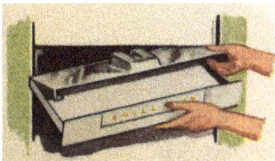

The addition of the Tri-Matic defrost for the G-85D model is added to the standard G-85 features of one pound butter tray, two plastic grid, two lever grid ice cube trays, meat drawer, two crisper drawers, chill tray and pantry door shelves.

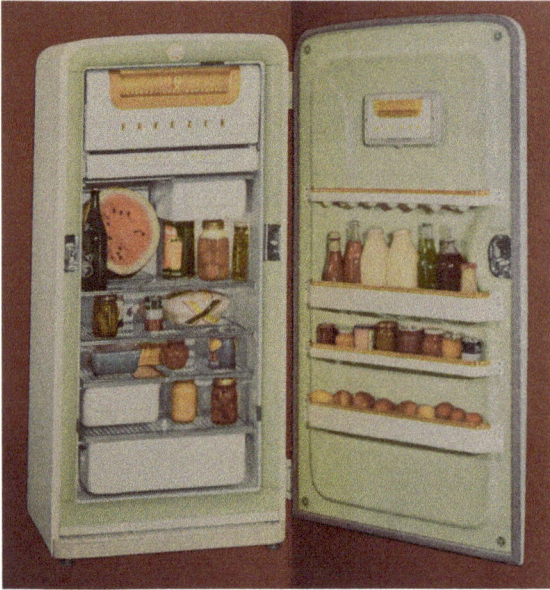

Model G-93 Household Refrigerator

G-93

9.3 cubic feet of food storage features a green breaker strip, green door liner and all white interior. A full length interior is set to luxury with gold trim. The three pantry door trays, an egg tray and butter keeper finish the fully loaded door options. Three Stainless steel full width shelves are joined with two porcelain-enamel crispers, one full width and another half-size, both with ribbed glass covers, for a total of 18.3 square feet of shelf area. The large frozen food area holds up to 51 pounds of food. A full width chill tray, covered meat drawer and four large lever release ice cube trays that make 64 cubes or 7.6 pounds of ice. Dimensions are listed in rear of book.

G-93 has four lever release ice cube trays, a one pound butter tray, light cover, chill tray with defrost cover, meat drawer, one and a half crispers drawers and pantry-door trays.

G-93D

With a white exterior, color coded handle, full spring-fresh green color interior, white porcelain-enameled crispers and drawers, stainless steel shelves, and gold trim accents, this refrigerator was the most colorful of the series. The addition of the 'Tri-matic" defrosting and chill tray insert brought the total weight of this model to 322 pounds. Shelving layout and crisper drawer options remained the same as the G-93. Suggested installed price for this unit in 1952 was $514.75 ($5,865.81 February 2023) Dimensions are listed in back of book.

G-93D is the luxury model for the 1952 sales year. Four lever release ice cube trays, one and a half crisper drawers, meat drawer, one pound butter tray, Diiufe-O-Lite, Defrost tray, Decorator door insert, plus tri-matic defrost.

Room
for
Everything
but
Doubt!

International
Harvester

WORLD'S
LEADING
FREEZERS

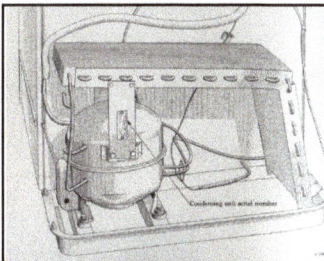

INTERNATIONAL
HARVESTER

L-12, L-16, L-20 Freezer
Serial number locations -
Left; Condensing unit number
on bracket above 'TightWad'
compressor.
Right; Cabinet Serial number on
cabinet unit, behind condenser
cover.

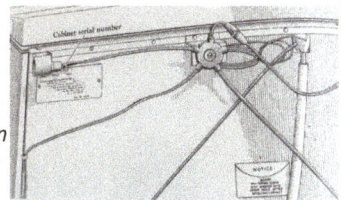

1953 L- series
L-7, L-12, L-16 and L-20 Freezers

These chest freezers were introduced for the 1953 sales year. The styling reflected the matching refrigerators, streamlined and modern for the era. Trimwork reflective of the shadowline styling, interior color was spring fresh green and gold trim with a extended L-shaped handle, these freezers included warning lights, interior lights, automatic temperature control, and baskets as standard items.

Counter balanced lids continue to be a selling feature, as well as the "tightwad" compressor.

Serial numbers were located on the rear bottom left for L-7 and on the interior of the condensing housing unit on the upper left for L-12, L-16 and L-20.

Model L-7 Freezer

A-30139

A-30140

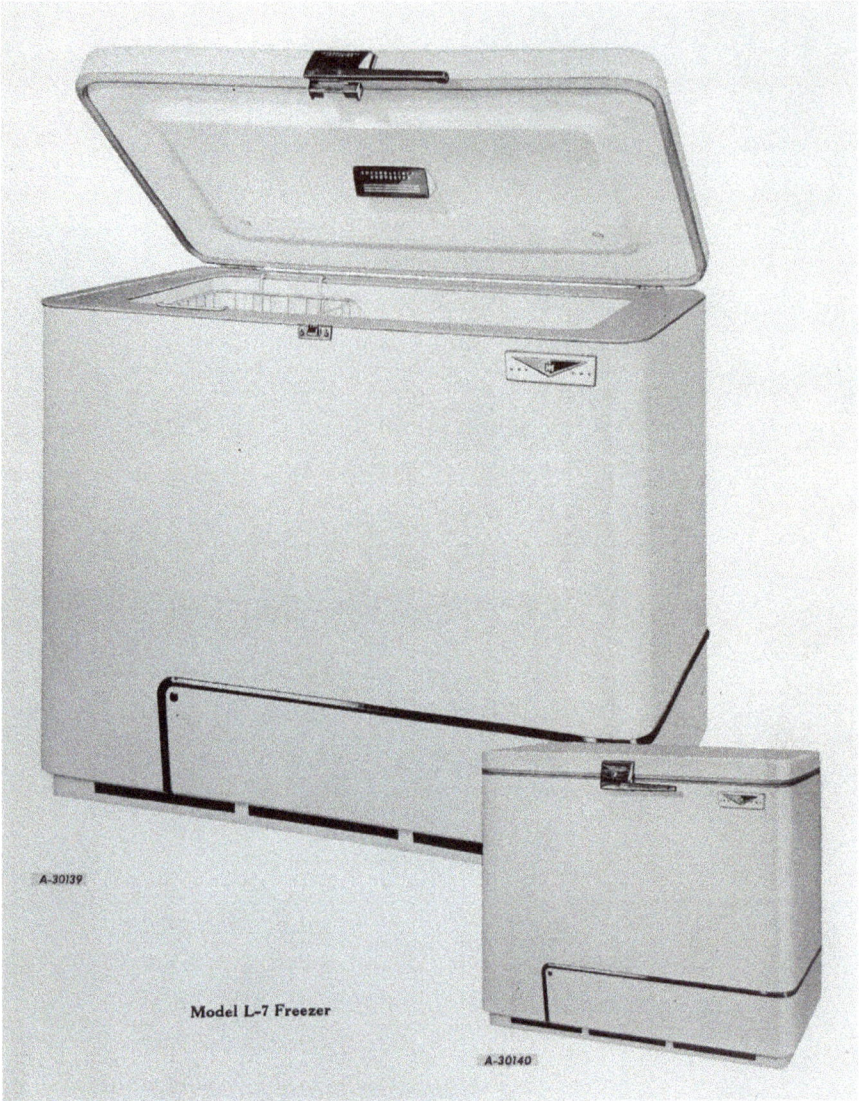

One basket and one divider were standard with the purchase of every L-7 Freezer.

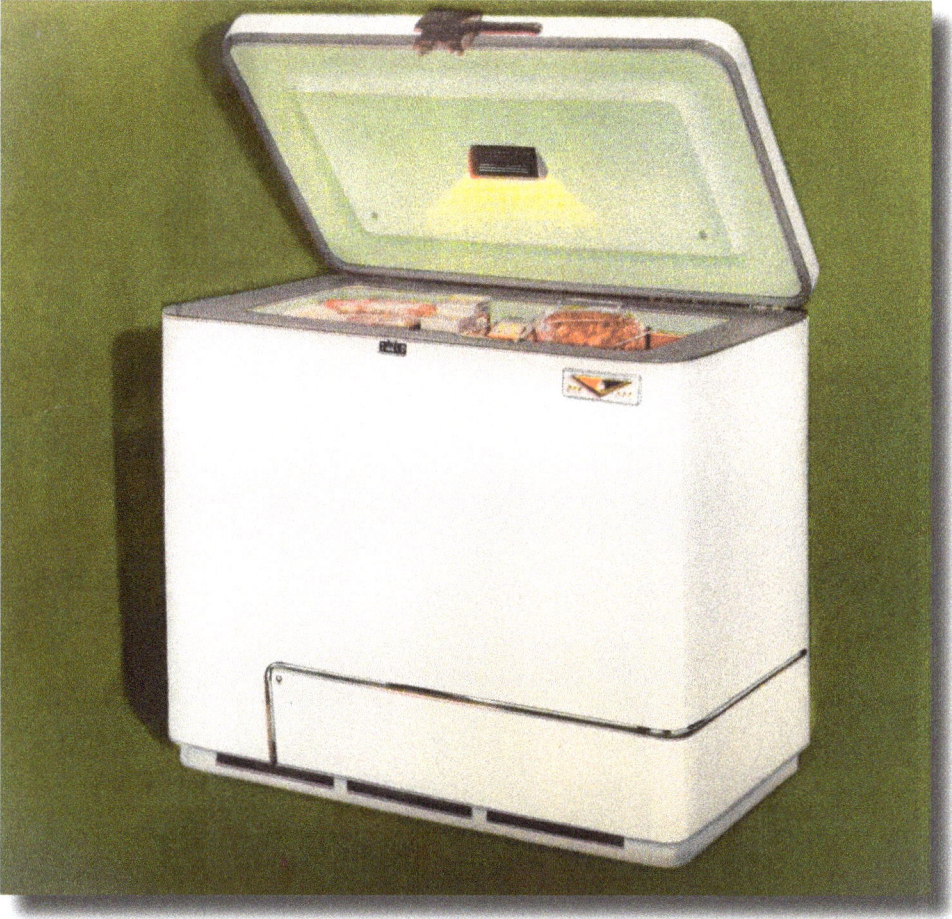

Location of Serial Number on L-7; bottom left on rear of cabinet.

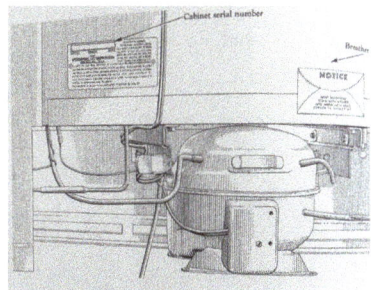

L-7

The smallest freezer for 1953 at 7 cubic foot size, came with one basket and one divider, non-locking handle, and could store 245 pounds of food. Interior is spring green with light integrated in lid. Four and a half inches of fiberglass insulation and 63 feet of coil keeps the interior cold.

Dimensions are 44" wide, 36" tall, 27" deep, 287 pounds, sized for inclusion in the kitchen. Serial number is located at the rear of the unit, on the warranty plate.

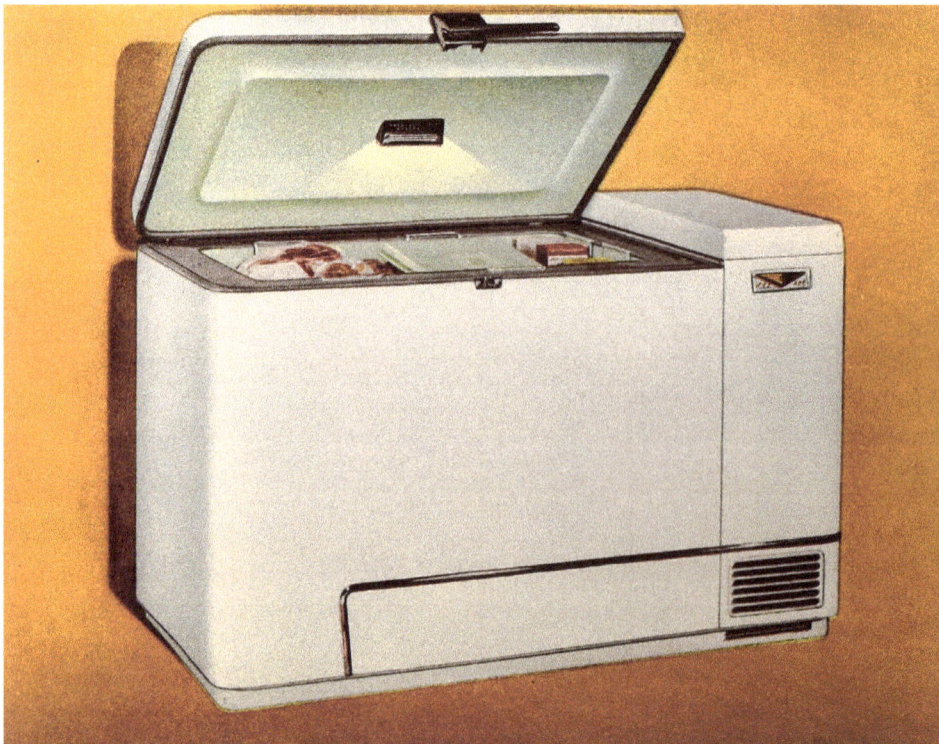

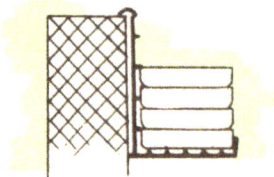

Two baskets, two dividers, two utility shelves, a hostess tray and one ice cube container were standard accessories for this L-12 Freezer. 1953 is the only year that the hostess tray and ice cube container were offered with purchase of freezer from International Harvester.

L-12

Radiant in spring green interior and interior light, the freezer arrived with a standard locking handle, two utility shelves, a hostess tray, ice cube container, two dividers and two baskets, holding 388.5 pounds of food, in 11.1 cubic feet. 29" deep, 36" tall, and 58" long, weighing 406 pounds.

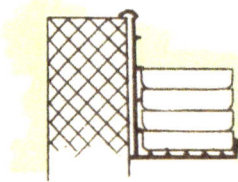

Three utility shelves, three baskets, two dividers, a hostess tray and a ice cube container were provided with every new L-16 freezer in 1953.

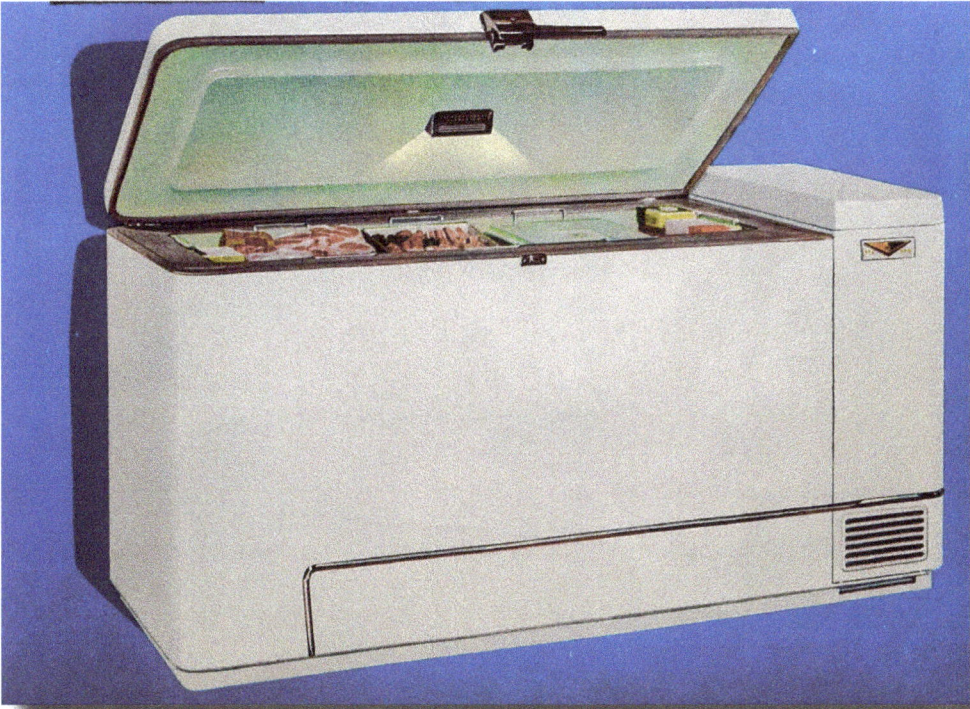

L-16

505 pounds of freezer to move, the L-16 featured storage for 560 pounds of food, in a 16 cubic foot area.

Arriving with three baskets, three utility shelves, two dividers, a hostess tray, ice cube container, and locking handle, this freezer fit into a 73 ¼" wide 29" deep space. At 36" high, 64 3/8" of clearance above was needed to open the freezer. Powered by a 1/3 horsepower Tight-Wad compressor, 4 1/2" of insulation keeps food frozen.

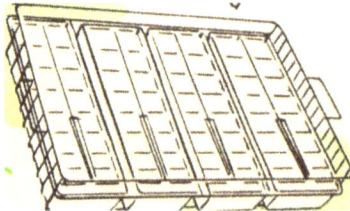

One hostess tray, a ice cube container, two dividers,
three stainless steel baskets, four utility shelves, an ice
cube rack with four lever style ice cube trays, all arrived
with the purchase of this L-20 freezer in 1953.

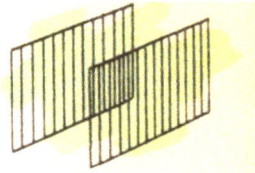

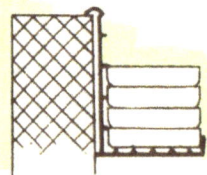

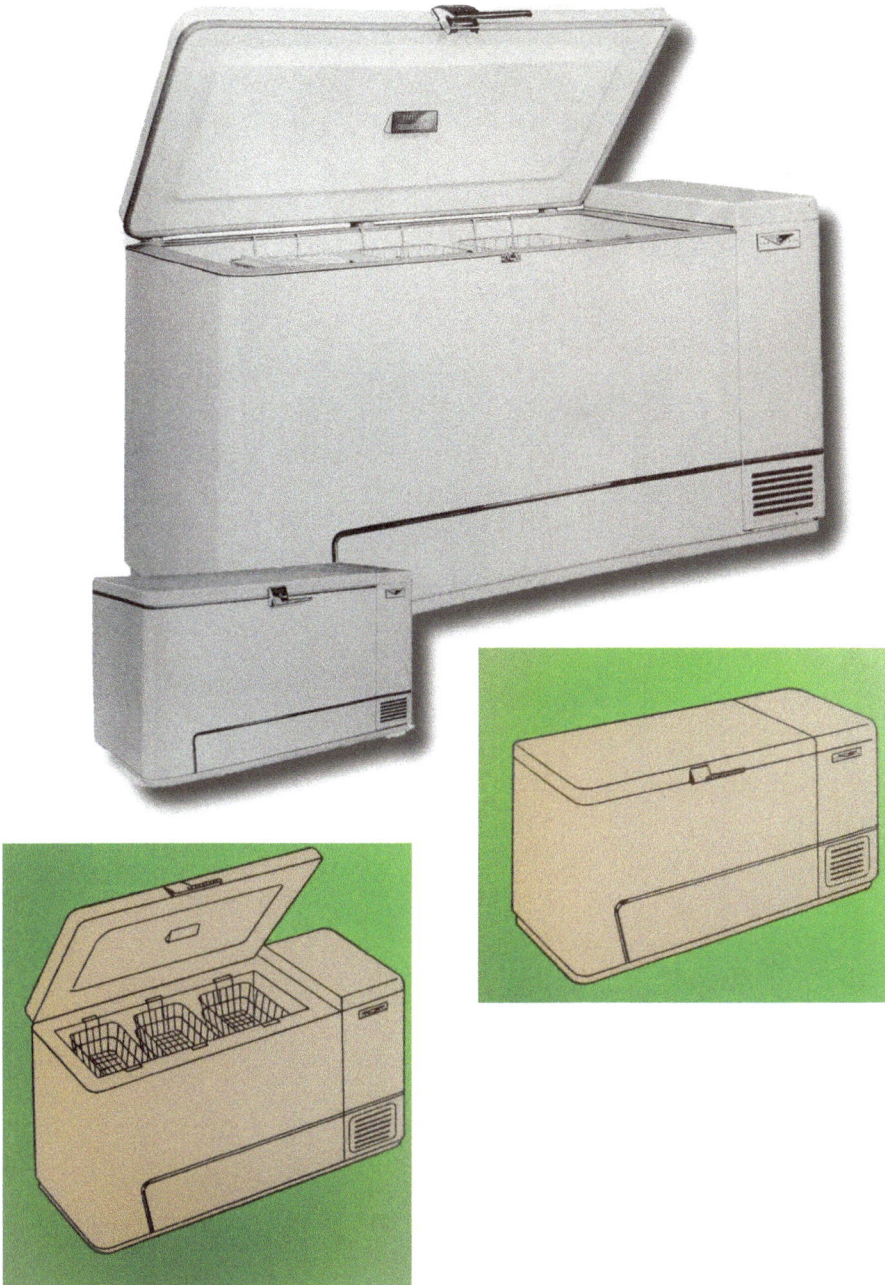

L-20

A behemoth of a unit at 20 cubic foot area, holds 700 pounds of food with spring green interior and dome style interior light. Weighing almost as much empty at about 530 pounds, this freezer arrived with three baskets, four utility shelves, two dividers, a hostess tray, ice cube container, an ice cube rack with four lever style ice cube trays, and a locking handle. Dimensions are 36" high, 29" deep, 73 ¼" long. Powered with 1/3 horsepower Tight-Wad compressor.

INTERNATIONAL HARVESTER VERTICAL FREEZERS

OWNER'S MANUAL

IH
INTERNATIONAL
HARVESTER

MODEL NOS. L-9-V AND L-14-V

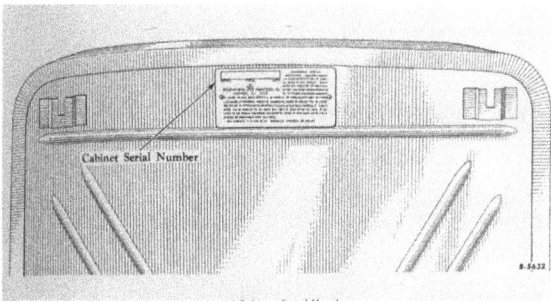

Cabinet Serial Number

Serial Numbers for the upright freezers are located on the warranty plate, at top on rear of cabinet for each model.

The condensing units are different locations.
L-9-V on side of coil cover.
L-14-V on the mount of the condensing unit.

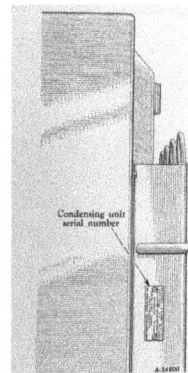

Condensing unit serial number

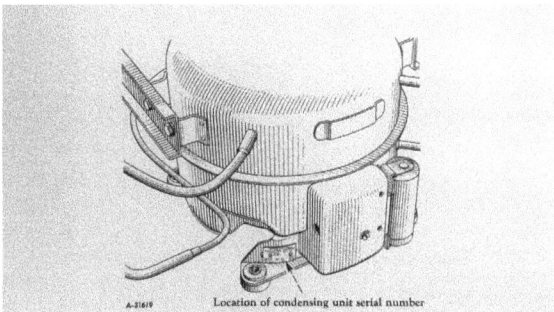

Location of condensing unit serial number

L-9-V and L-14-V

These upright freezers were the first upright freezers that harvester offered, releasing them for sale mid- 1953, seven years after the first announcement of the idea. Serial numbers for the cabinet are found in the upper middle of the rear of the unit, between the wall spacing mounts. The condenser serial number is found on the side of the coils. (see photos)

Features of the upright freezer included refrigerated shelves, a preset temperature control for 0° average cabinet temperature, R-22 refrigerant, standard 5 year warranty, inside door light and exterior operating light.

NEW IH Upright Freezer BY INTERNATIONAL HARVESTER MODEL L-9-V Capacity, 8.7 cubic ft. Holds 305 lbs

L-9-V

From the outside looks like a regular refrigerator cabinet, complete with Shadowline styling and upside down 7. Cabinet dimensions are 59 5/8" high, 29 3/8" deep, 29 ½" wide, and weighs 310 pounds. Inside, the spring fresh green interior surrounds three refrigerated shelves, two sliding drawers and three pantry-door shelves that held small items like concentrated frozen juice or ice cream bars. Interior light in panel of door, and outside light-up emblem for peace of mind operation. 1/4 horse-power 'Tight-Wad' compressor.

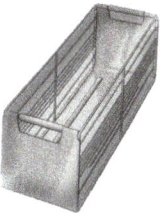

The L-9-V upright freezer arrived with two sliding gold colored storage drawers; holding 20 pounds each.

Cabinet front blends with L-Series left-hinge refrigerator, for a 1953 matching side by side freezer and refrigerator in the kitchen.

L-14-V

The exterior is streamlined, featuring the International Harvester refrigeration emblem and a latching, lockable door handle. The interior of this 14 cubic foot freezer has three refrigerated shelves, and one adjustable, non-cooled shelf for a total of four shelves. Three sliding drawer shelves hold 20 pounds of food each, and three fruit juice dispensers.

Exterior measurements of this freezer are 68 ½" tall, 30 5/8" deep, 29 ¾" wide and weighs 375 pounds.

A-31400

A-31401

NEW International Harvester
Upright Freezer
BY INTERNATIONAL HARVESTER
MODEL L-14-V
Capacity, 14 cubic ft. Holds 490 lbs.

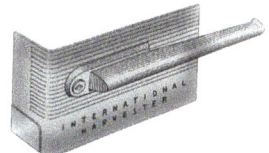

The L-14-V upright freezer arrived with three sliding gold color drawers, three fruit juice can dispensers, a lighted emblem and a lockable handle.

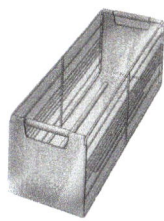

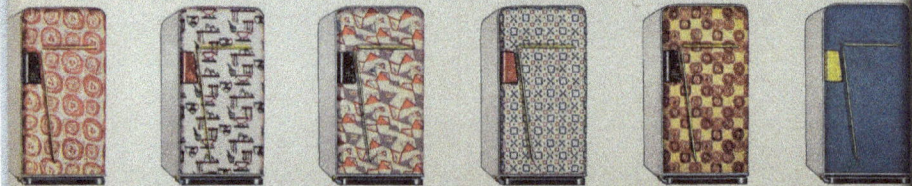

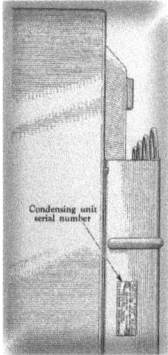

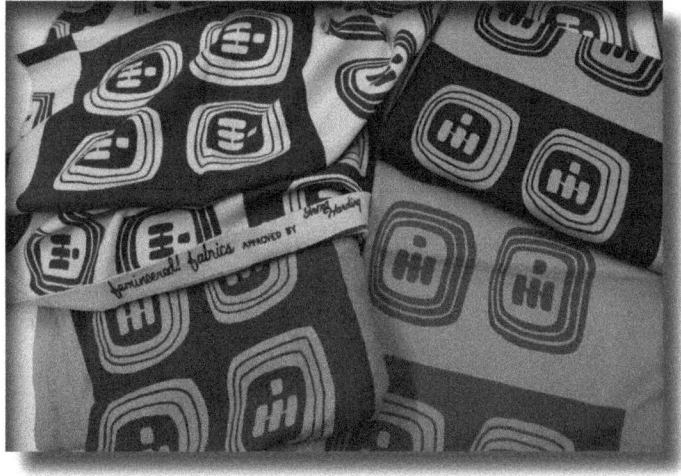

Identification numbers for the L-Series refrigerators:

Condensing unit is located on side of coil cover. (above)

Serial Number located on top rear of cabinet, as part of the warranty plate. (right)

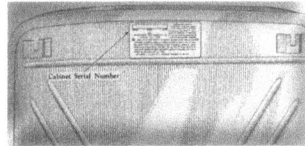

1953 L-Series Refrigerators

The complete L-series line for 1953 can be grouped into three categories:
Apartment sized, L-74 and L-82
Regular sized; L-84, L-84-D, L-84DM L-85-D
Luxury sized; L-100-D, L-100-DS, L-103, L-103-S, L-105-D, L-105-DS L-105-DM and L-105-DMS
Utilizing the traditional "Shadowline Styling," the exterior of the refrigerators feature a white cabinet, black toe kick, an embossed emblem in gold and black accents for smooth cleaning and a new, downward pull handle in that looks like an upside-down '7'. The option to cover the front of the refrigerator in material to complement the space on some of the models could be changed in about seven minutes. Remaining standard on all refrigerators is the front adjusters to level the unit as well as an integrated bottle opener in the door latch.
Seven different climates inside to meet each food need, finished out the lucky sevens feeling.
This Series was the first to offer an option to open the door by a foot pedal.

A-30180

A-30182

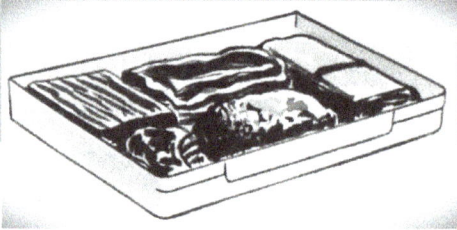

Two plastic grid ice cube trays and meat chill tray were standard accessories with the L-74 refrigerator.

L-74

An economical option, 7.4 cubic feet of storage featured the clear view freezer door, hinged at the top and in spring green, holds 35 pounds of food. Two plastic grid ice cube trays and 14.5 pounds of meat chill tray and bottle opener in door latch. The trim and door insert were the spring green with a white interior wall. Three change-able and adjustable zinc coated wire shelves with a chill tray offered multiple shelf arrangements.

278 pounds, and 28 1/2" deep, 24 7/8" wide, 54 3/8" tall.

Standard accessories for the L-82 were two plastic grid ice cube trays, a meat chill tray, and a one pound butter dish that fit inside the heated butter compartment.

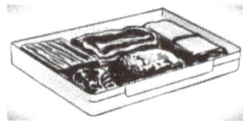

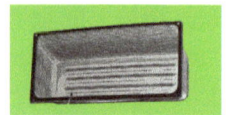

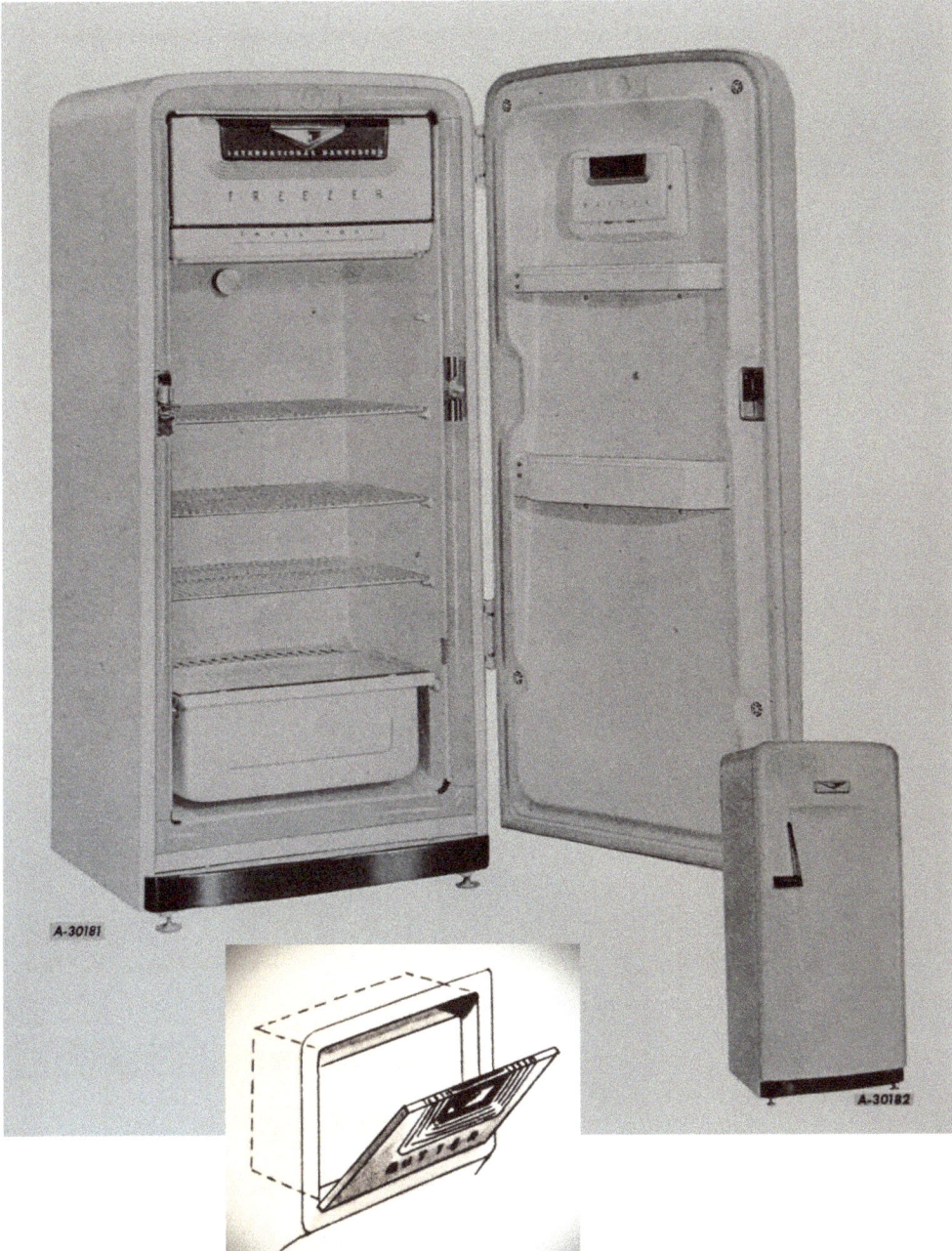

A-30181

A-30182

L-82

Arrived with a full spring green interior, white full width freezer door, hinged at the bottom also featured a full width crisper with glass cover. The in-door pantry featured two shelves, one for eggs and another for condiments and a butter keeper with spring-hinged door with butter tray. Three chrome plated shelves and meat chill tray with two plastic grid ice cube trays gave many shelf arrangements.
294 pounds, 28 1/2" deep, 24 7/8" wide, 54 3/8" tall.

Above: L-103 Refrigerator Serial Number 501-700. These first two hundred built featured a different interior pantry door than the ones built after Serial No. 701.

Covered meat drawer, full width crisper, butter tray, two plastic grid and two lever style ice cube trays were standard accessories for the L-103 refrigerator.

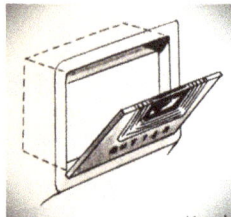

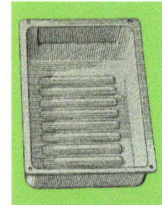

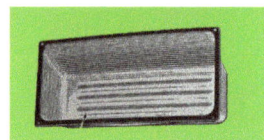

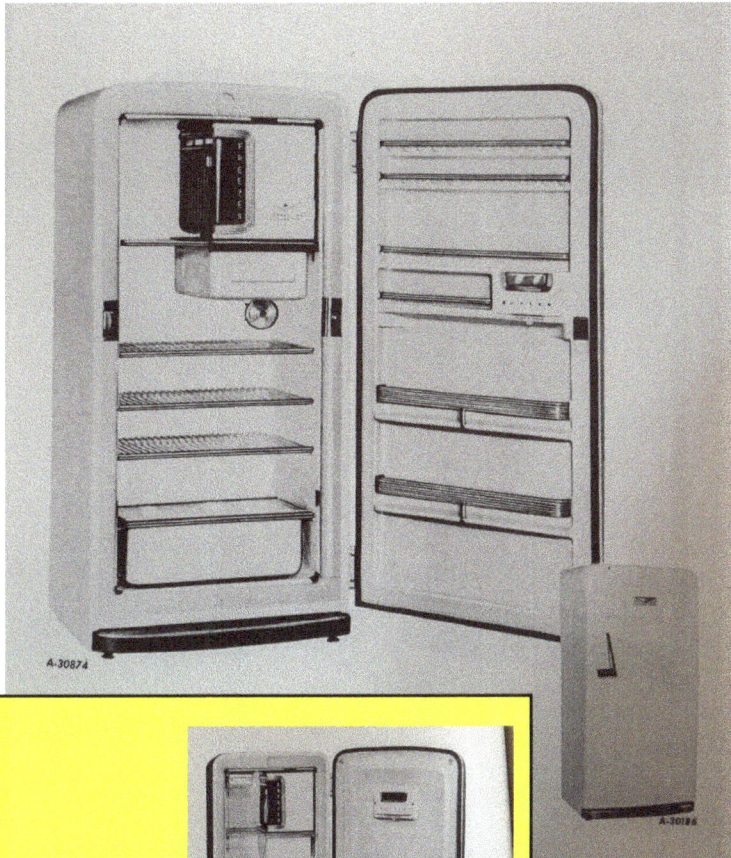

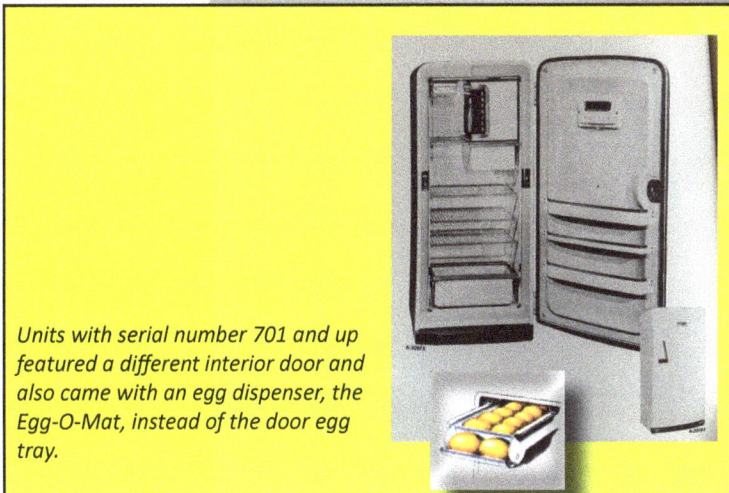

Units with serial number 701 and up featured a different interior door and also came with an egg dispenser, the Egg-O-Mat, instead of the door egg tray.

L-103

The largest economy version, with a U-style freezer, white interior, spring green trim and interior door. Two plastic grid and two shucker type ice cube trays were also included. Below the freezer is a covered meet tray, three and a half chrome plated shelves with gold trim, and the bottom featured a ribbed glass covered full width crisper.

Pantry door on units SN 501 to 700 featured two egg shelves, holding 24 eggs total, a upper and lower utility shelf, heated one pound butter keeper, and two bottle shelves.

Pantry door on units SN 701 and up featured three utility trays and heated one pound butter keeper. This model also came with the Egg-O-Matic.

Measurements are 29 1/2" wide, 29 3/8" deep, 59 3/4" tall, weighing 295 pounds.

Three plastic grid ice cube trays, a meat chill tray and a half width crisper were standard accessories with this L-84.

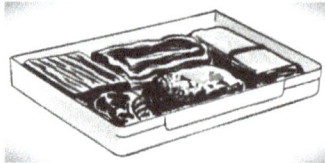

A-30183

Model L-84 Household Refrigerator

A-30186

L-84

A modest option for many families, it features a full width freezer with fold down shelf, chill tray, half shelf and half crisper. Two trays in the green backed interior door, one egg and one for condiments. The interior is white, with green trim to match. Three plastic grid ice cube trays finish the accessories for this unit. The three and a half chrome plated shelves hold 17.6 square feet of space.

Measurements are 29 3/8" deep, 29 1/2" wide, 59 3/4" tall and weighs 290 pounds.

A-30184

A-30186

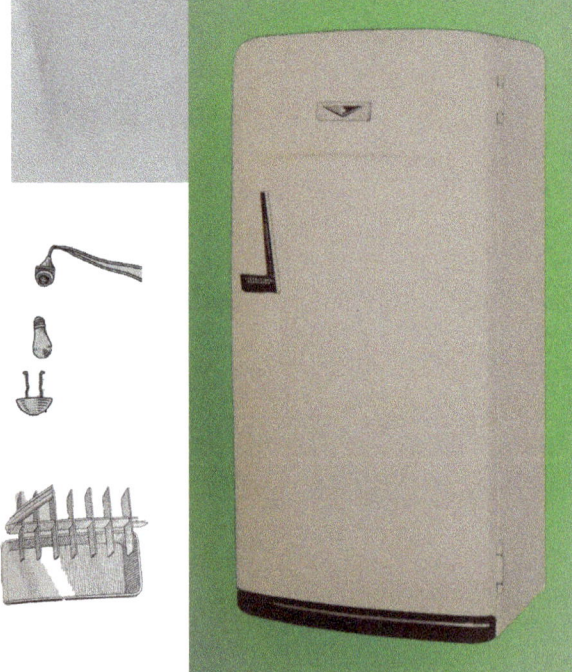

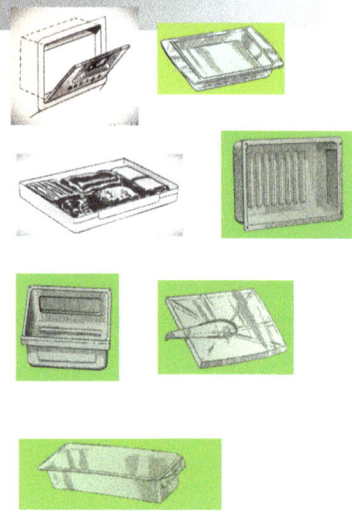

L-84-D and L-84-DM

Features like a fifteen minute push button defrost cycle and water catch, white interior, with an addition of the heated butter keeper and meat tray. 8.5 cubic foot, large, 50 pound storage freezer capacity chill tray and defrost tray, covered meat storage, and deep crisper drawer with ribbed glass cover. Three lever release ice cube trays, chrome plates shelves, two pantry door trays, magnetic bottle opener. Dimensions are 29 3/8" deep, 29 1/2" wide, 59 3/4" tall and weighs 300 pounds.

L-84DM

Decorator model, the option for material covering on the exterior, with foot pedal release. All Features of the L-84-D were the same for L-84-DM (photos on next page)

A-30870

A-30871

Outer door pan

A-30916

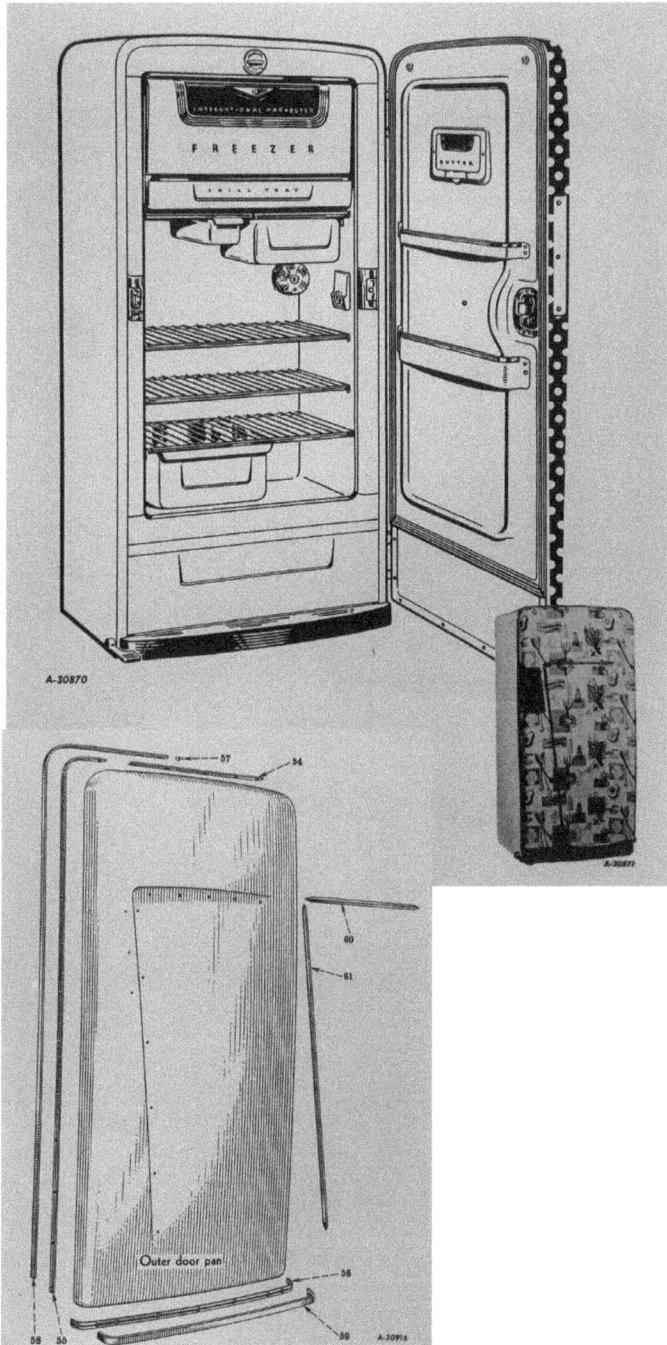

L-84-D and L-84-DM refrigerators arrived standard with heated one pound butter compartment and tray; chill tray, three lever release ice cube trays; water catch, covered meat drawer and covered crisper. Light was covered with a "Diffuse-O-Light" ; push button defrost.

L-85-DM also arrived with removable gold trim, push plate and lever door release action. A skilled person could change the door material in about seven minutes.

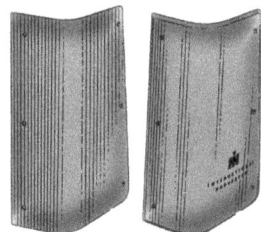

A-30185

A-30186

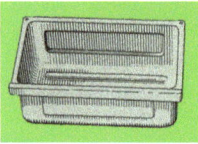

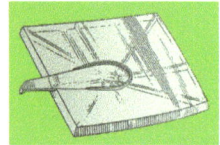

L-85-D arrived standard with heated one pound butter compartment and tray; chill tray, four lever release ice cube trays; water catch, covered meat drawer and two covered crispers. Light was covered with a "Diffuse-O-Light" ; push button defrost and wire evaporator rack inside the freezer.

L-85-D

Arrived with a spring fresh green inner liner, door and trim, push button defrost, twin crispers and meat drawer with one pound butter keeper, egg, utility and bottle trays. Two chrome plated wire shelves, four lever release ice cube trays, full width chill tray and freezer complete the 8.5 cubic foot refrigerator.

Magnetic bottle opener in door latch, this unit is feature packed.

Measurements are 29 3/8" deep, 29 1/2" wide, 59 3/4" tall and weighs 309 pounds.

L-100-D Serial Numbers 501 to 700 feature a different Pantry door. All a refrig-
erators arrived with heated one pound butter compartment and tray; chill tray,
four lever release ice cube trays; water catch, covered meat drawer and full width
covered crisper. Light was covered with a "Diffuse-O-Light" ; push button defrost
and wire evaporator rack inside the freezer.

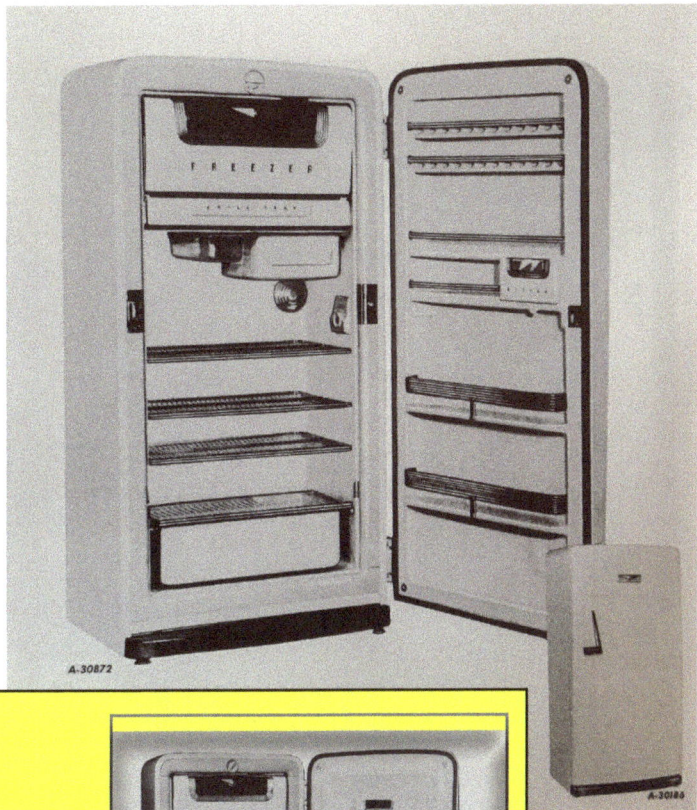

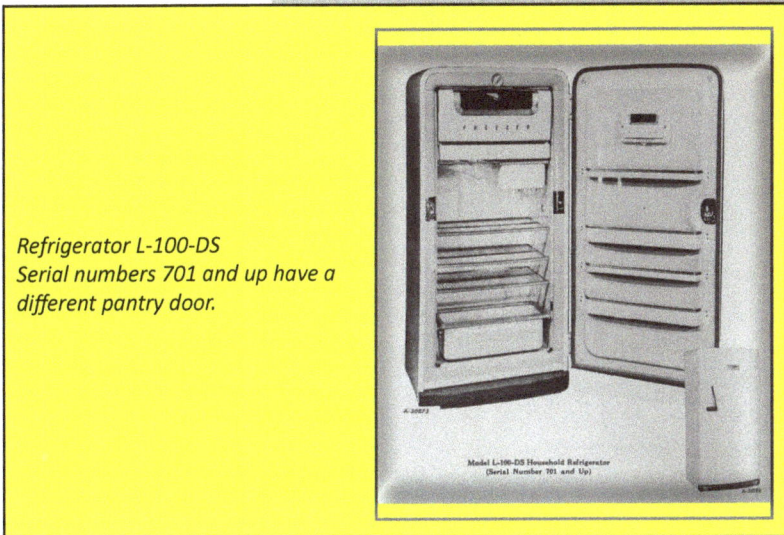

*Refrigerator L-100-DS
Serial numbers 701 and up have a
different pantry door.*

L-100-D

A large 10 cubic foot unit, and a total spring green interior, utilizes the whole refrigerator for storage. Full width freezer, chill tray and crisper with glass lid as well as a meat drawer and push button defrost. Stainless steel wire shelves with trim, four lever action ice cube trays and a water collection tray.

Units with serial number 501-700 inner door has two egg trays, two utility trays, a heated butter keeper, and two bottle shelves.

Units with serial number 701 and up inner door has three utility trays, one egg tray and heated one pound butter keeper. All other details remain built the same.

Dimensions are 29 3/8" deep, 29 1/2" wide, 59 3/4" tall and weighs 311 pounds.

Refrigerator L-105-D and L-105-DM serial number 701 and up were stamped as L-105-DS and L-105-DMS to reflect the inner door liner change. All other details remained the same.

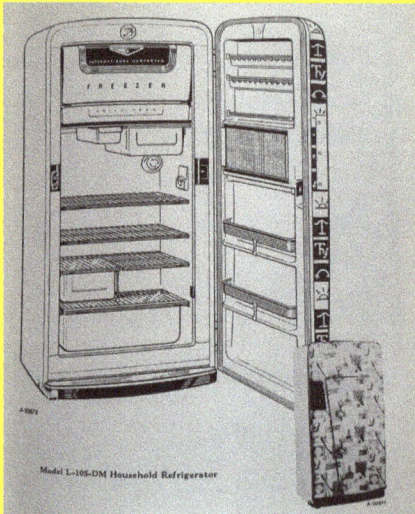

Model L-105-DM Household Refrigerator

Model L-105-DMS Household Refrigerator
(Serial Number 701 and Up)

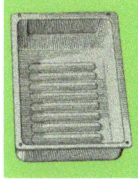

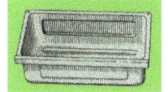

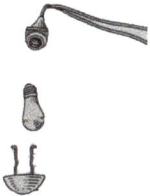

All L-105-D models arrived with four lever type ice cube trays, wire shelf inside freezer, chill tray, defrost cover, water defrost container, meat drawer, full width crisper, half width crisper, one pound butter tray and Diffuse-O-Light cover.

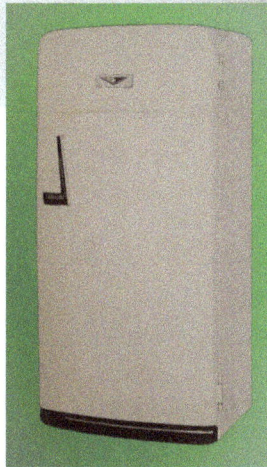

L-105-D and L-105-DM

These two models combined every previous feature with three stainless steel wire shelves with trim, wide width freezer, chill tray, meat drawer, one full and one half-width crisper and push button defrost with water collection container.

Door on units SN 501-700 became a super pantry, holding 24 eggs in two trays, a utility tray for small jars, two more trays for bottles, and sliding glass panels to keep butter, cheese, and bacon behind.

Door on unit SN 701 and up feature a egg tray, utility tray, two bottle trays and a heated one pound butter keeper with butter tray.

All of this was framed with spring fresh green interior and gold anodized retainer guards. 10 cubic foot of storage, and 19.6 square foot of shelf storage.

Dimensions are 29 3/8" deep, 29 1/2" wide, 59 3/4" tall and weighs 330 pounds.

L-105-DM is the decorator model, with foot lever release and push plate, all other details are the same as the L-105-D model.

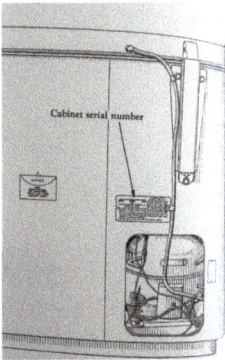

Cabinet Serial Number for deep freezers can be found on the warranty tag, at rear above the condenser.

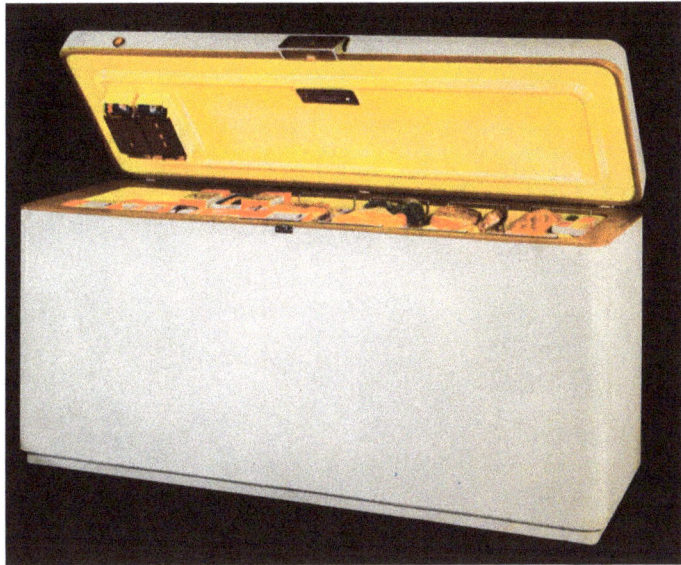

1954 M Series Freezer
M-7, M-12, M-16, M-20

The freezers introduced for sale in 1954 features the new sunshine yellow interior with brown seals, mimicking the bronze details found on the outside of the cabinet. The warning light was moved to the front left of the latch on the lid and the freezer no longer had a removable condenser cover. The space above the motor was integrated into part of the freezing area, as we are familiar with in 2023. Serial number plates are all found to the rear of the cabinet above the condenser motor, towards the bottom. A pattern was also included for adding decorative fabric to the top of the freezer, to match the refrigerator décor on every model except M-7.

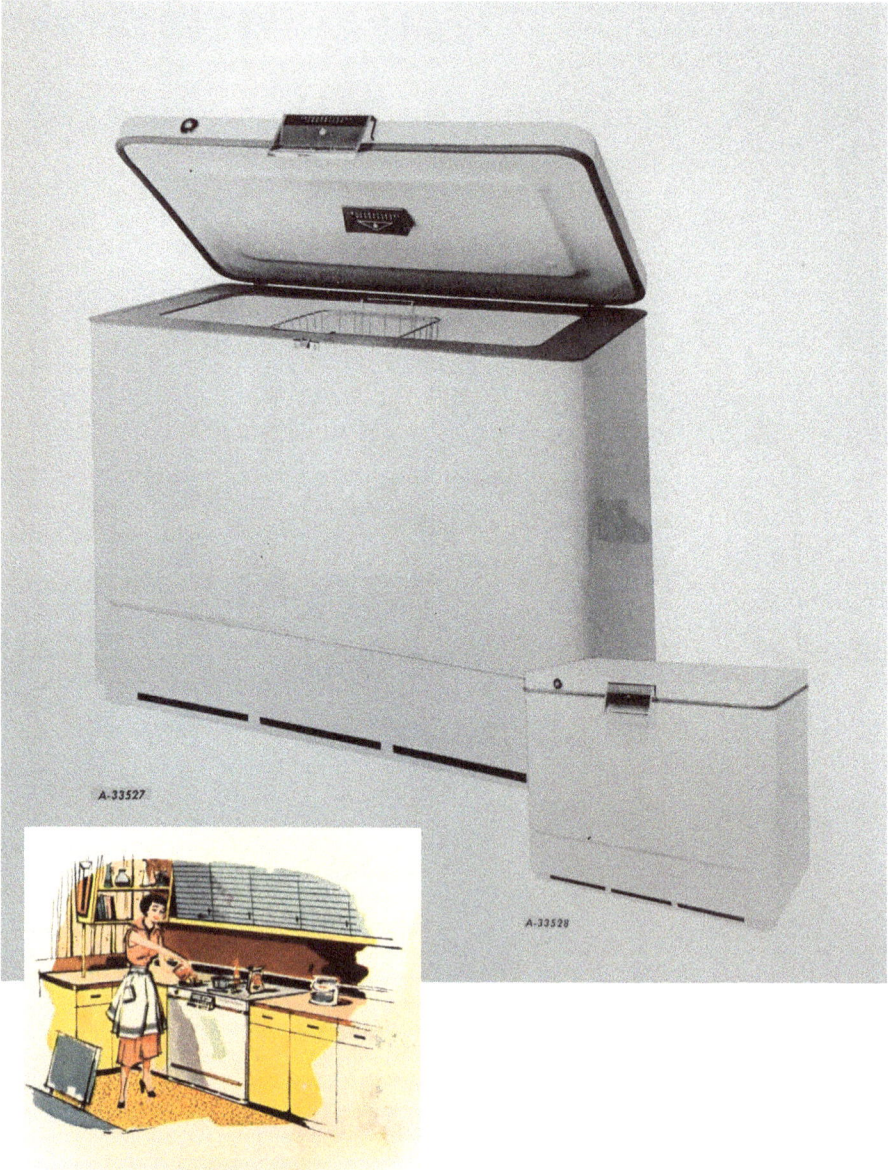

M-7

The smallest of the units, the 7 cubic foot model was designed to feature in any home kitchen. A toe recess and decorative bronze trim matched the bronze lid latch and interior light cover. A wire divider and one wire basket were also supplied with the unit. Measuring 36" tall by 27" deep and 44" wide, this freezer was powered by a 1/6 horsepower motor, and weighed 264 pounds.

M-7 Freezer came with one basket, and one divider.

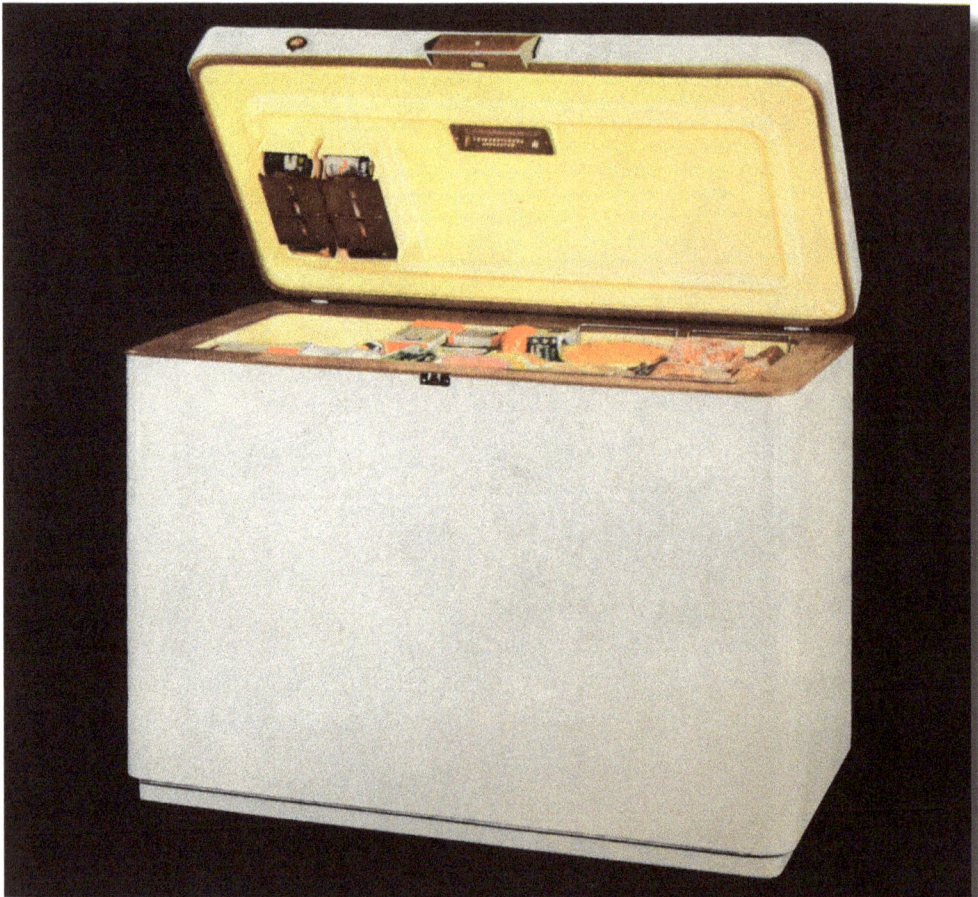

M-12

Two additional features in this 12 cubic foot freezer are the pattern for a decorator lid and the addition of two juice can dispensers. The bronze trim on the front was removed for these larger sized freezers, leaving the toe kick area and light-up emblem or warning light on the front left of the locking latch on the lid. The 366 pound freezer came with two baskets, and one wire separator. Measurements are 26 1/2" deep, 37" tall, 54 ¼" long, powered by a ¼ horsepower motor.

A-33529

Model M-12 Freezer

A-33530

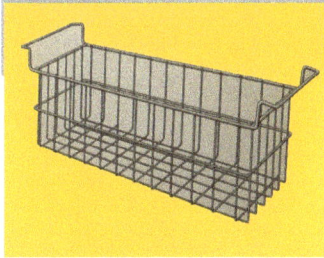

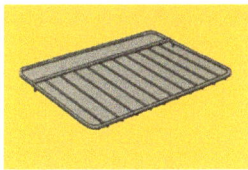

M-12 arrived with two baskets, one wire seperator, and two juice dispensers.

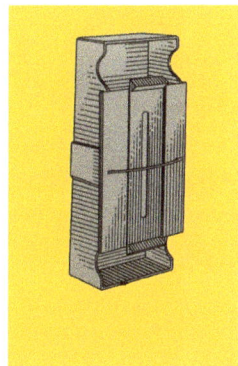

A-33531

A-33532

M-16

Three baskets, two dividers, and a dual juice can dispenser could be found inside the 16 cubic foot freezer. The decorator pattern and all other features remained the same as the smaller unit. Using a 1/3 horsepower motor, the 437-pound unit measured 26 ½" deep, 37" tall, and 65 3/8" wide.

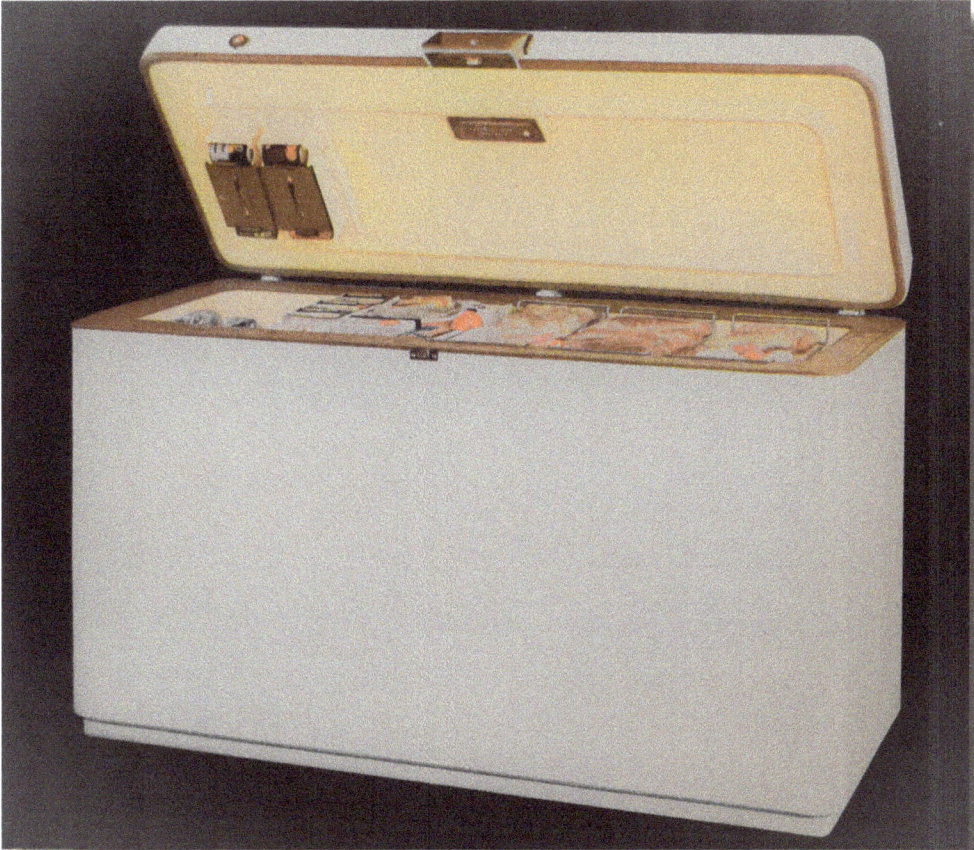

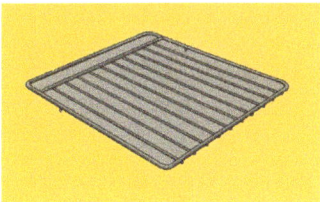

M-16 has two juice dispensers, two dividers and three baskets.

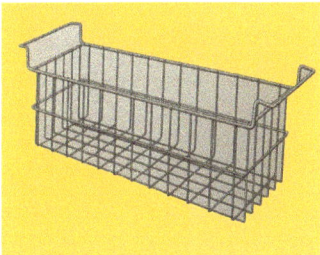

M-20
Holding 700 pounds of food inside this 78 ½" wide, 26 ½" deep and 37" tall freezer using four baskets, three dividers, and a dual juice can dispenser, this 20 cubic foot freezer weighed 460 pounds. Powered by a 1/3 horsepower motor, the attractive bronze locking latch and lighted International Harvester emblem complemented this M-series freezer.

A-33533

Model M-20 Freezer

A-33534

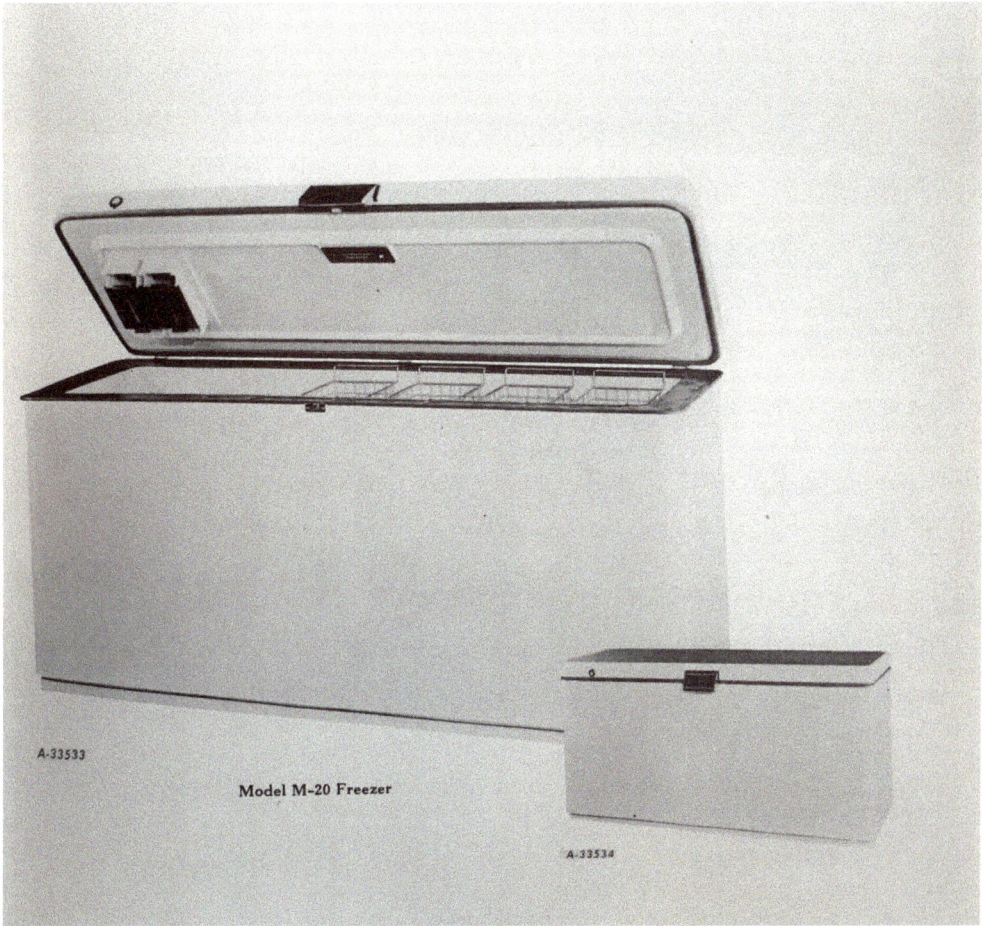

M-20 came with four baskets, three dividers, two juice dispensers and one decorator pattern.

Room for this and everything else in the all-new IH refrigerators

Plenty of space for the tall, the small, the bulky . . . for everything you eat . . . and all within easy reach! All the new features you've wanted, too—a 50-pound freezer, push-button automatic defrosting, Super Pantry-Dor, adjustable and roll-out shelves. Plus the only refrigerator door you can decorate to match your kitchen.

Choose from 7 new models at your IH dealer's—his name is in the Yellow Pages of your phone book.

International Harvester

Extra room for summer foods... zero cold to keep them fresh... in the all-new IH freezers

Freeze these tasty foods right now—while prices are low, quality high. Enjoy them any time—freshness and flavor are captured in your new IH freezer. International Harvester gives you more space for any kind of food—meat, poultry, produce . . . the *correct* cold to keep it safe . . . extra convenience with deep, wide door shelves and roomy, roll-out baskets. And IH is the only freezer you can decorate—chest or upright. Choose from 7 models, from 245 to 700 pounds capacity.

International Harvester

1954 Magazine advertising for the M-Series refrigerator and freezer. These two advertisements were found in a magazine on different pages. TC

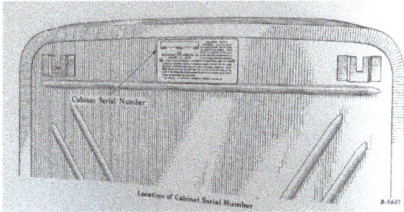

Serial number for the M-Series vertical freezers can be found on the warranty plate located top middle at rear of cabinet.

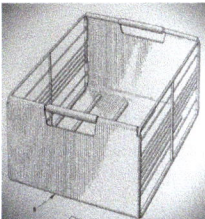

MV-9, MV-15, MV-19

The lineup of vertical freezers saw an addition for 1953, the 19 cubic foot freezer. The three models all had bronze trays on the inside door, interior lighting (on inside of door) and bronze fronted roll out baskets at the bottom. The serial number tag could be found at the upper rear of the cabinet, between the wall spacer brackets.

MV-9 arrived with two roll-out baskets

MV-9

Designed to complement the Left- hand opening refrigerators, in both style and size, it has three refrigerated shelves for fast freezing, two pantry door trays, and two roll out baskets and an interior light in the door.

A warning light, located in the IH emblem on the decorator trim of the door matched the M-85, M-104 or M-105 series refrigerators. A pattern for decorating and a non-locking door handle completes the look for this 8.7 cubic foot, ¼ horsepower, 298 pound upright freezer. Cabinet dimensions are 29" wide, 29" deep, 59 ½" tall.

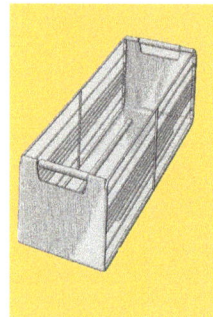

MV-15 has three roll-out baskets

MV-15

A lockable latch on the 14.7 cubic foot freezer prevented small children from being able to eat all the ice cream bars stored in one of the four pantry door trays. Three refrigerated shelves and one adjustable shelf created a lot of room to store 514 pounds of pie and ice cream. Three roll out baskets along the bottom of the freezer kept the fruit accessible when it was time to make more pie.

Powered by a 1/3 horsepower motor, the cabinet measures 29 ½" wide, 29 ½" deep and 68 ½" tall, and weighed 390 pounds, about how much I would weigh if I ate only pie and ice cream. A pattern was provided, to cut material that featured pie or ice cream, so the front of the freezer could match what was inside.

MV-19

Holding 665 pounds of food inside 18.9 cubic feet, this freezer measures 30" deep, 36 ½" wide, 72 ¼" tall on the outside. Weighing 451 pounds, it is probably still working in 2023 where it was installed in 1954. Bronze trim and sunshine yellow interior, with a white exterior, would complement a bright sunny kitchen in any time period, especially when the front of the freezer can be changed to match the décor chosen. Three hanging juice dispensers on the inside with four shelves (one adjustable, three refrigerated) to store food, and four bronze fronted rolling baskets were kept frozen by a 1/3 horsepower motor maintaining 0°F storage. The only freezers able to match the décor were built by International Harvester.

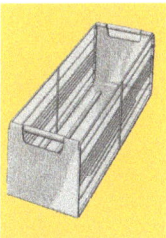

MV-19 arrived with four roll-out drawers and three juice dispensers.

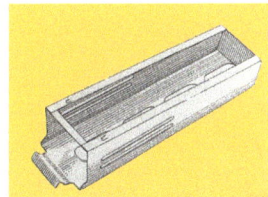

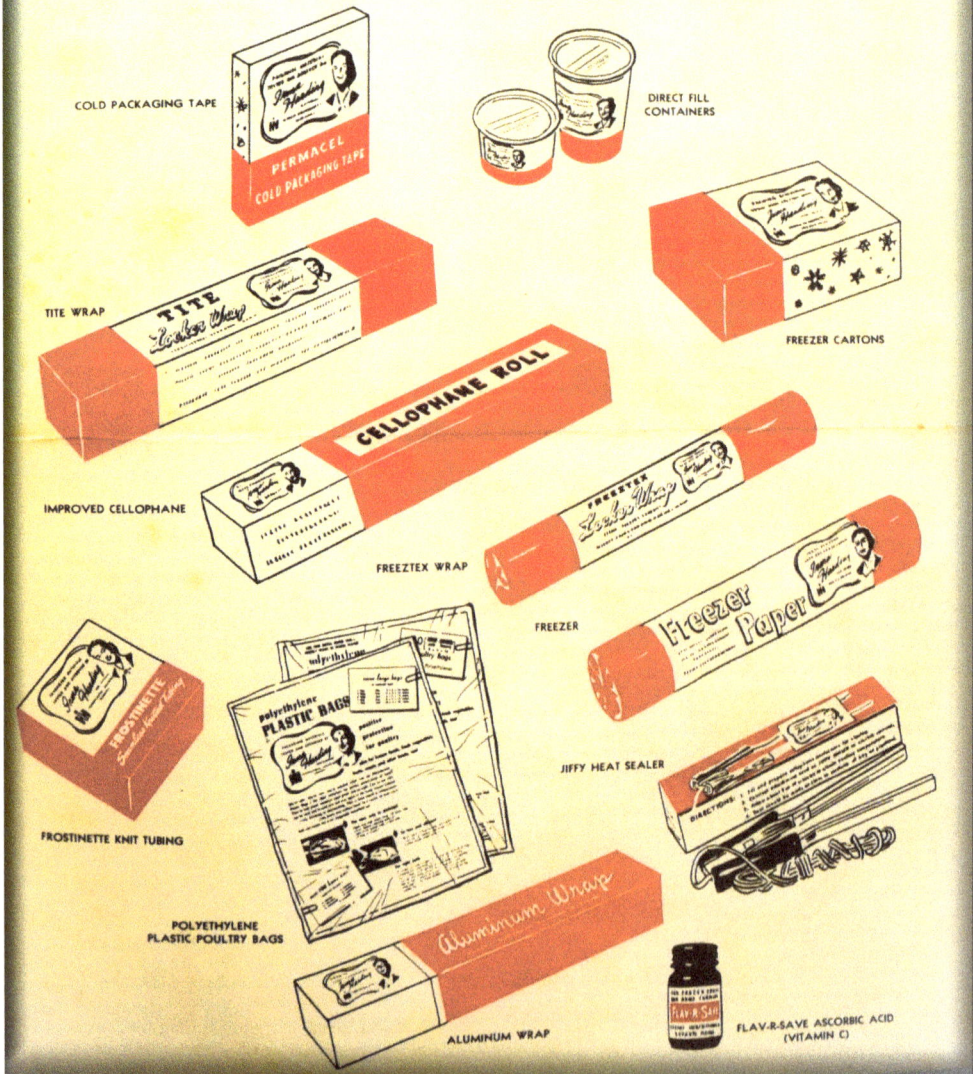

Irma Harding began making appearances in 1949, as a marketing image for the refrigeration line. Led by the women of the International Harvester test kitchen and home economics, these ladies were instrumental in providing quality products and classes on how to freeze and store foods. Refrigeration was a new concept and it requires a different storage method than in the past with the ice blocks. Food safety and contamination was always important to Harvester. Early education began as part of the Service Bureau that started in 1912, offering demonstrations on how to safely prepare and store food by various methods by canning, dehydrating, or freezing as well as when to harvest and how to grow healthy gardens.

It was a natural extension to offer packaging materials as part of the line. Bager Products Co. of Neenah, Wisconsin and KVP of Kalamazoo, Michigan were the two suppliers of these items.

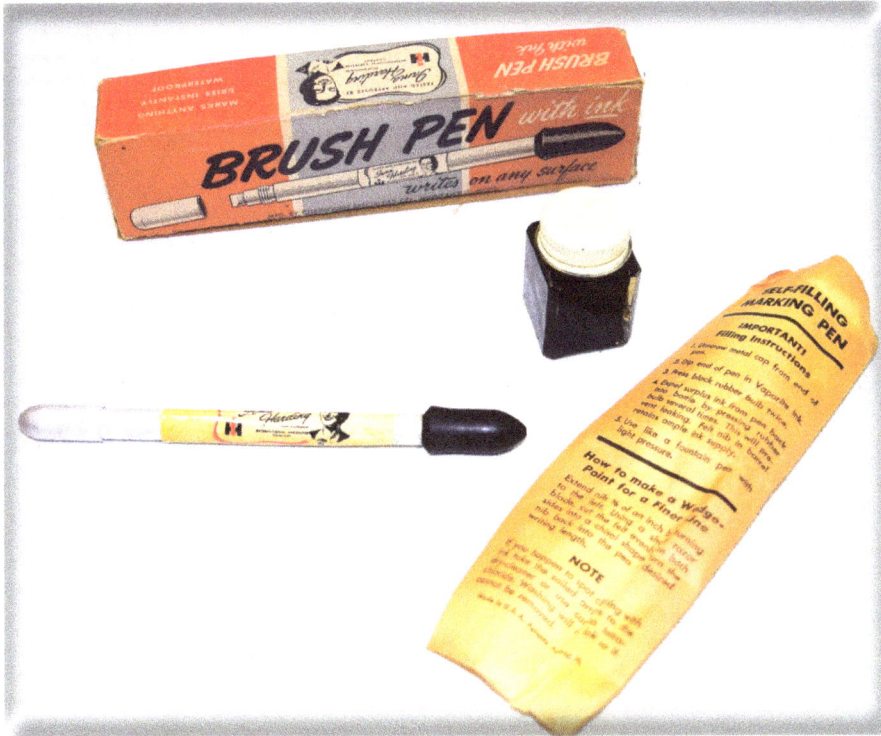

1954 M-Series Refrigerators

M-75, M-75-L, M-82, M-82-L
M-85, M-85-D, M-85-L, M-85-DL
M-104, M-105-D, M-105DX

A beautiful yellow porcelain interior in a satin smooth finish complemented the bronze trim for all models in this M-series. The first series that offered left hand hinges, the L designates a left-hand hinge option for select models.
All new door front styling, every model offered could be left white or color matched to the kitchen by adding fabrics to the upper or lower sections of the door front. This was the only manufacturer to offer the fabric matching option.

Serial tag remains located at top middle, between two brackets for the wall spacer bolts.

M-75, M-75-L

Designed for a small space, this 7.4 cubic foot model features three adjustable zinc plated shelves, a full width clear view freezer with 35 pound capacity, a chill tray holding 14 pounds of steak below and it also came with two 14-cube plastic grid ice trays.

Overall dimensions for this model are 24 3/8 " wide, 28" deep and 54 3/8" high.

M-75 and M-75-L (Left-hand hinge) arrived with two plastic grid trays and one chill tray.

A covered meat tray could be special ordered for this model.

M-82, M-82-L
Three Chrome plated shelves, a sixteen-quart glass covered crisper, two pantry door trays, and a in-door butter well combine to offer 8.2 cubic foot of storage in this small space, big luxury refrigerator. A full width, 35-pound freezer with bronze trim came with 2 ice cube trays, making 28 cubes.
Overall dimensions for this model are 24 3/8 " wide, 28" deep and 54 3/8" high.

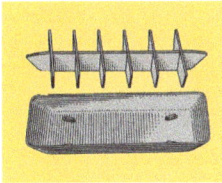

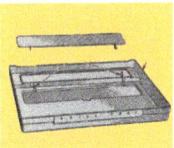

M-82 and M-82L has a full width crisper, one pound butter tray, chill tray, two plastic grid ice cube trays.

M-85, M-85-L

The entry level refrigerator in the large cabinet models, this unit offered three chrome plated shelves, a defrost water container on the side wall, a meat drawer of coated porcelain enamel steel, a half-width crisper with ribbed glass shelf and wire shelf, and the large full size fifty-pound capacity freezer. Three 14-cube ice trays, a egg rack and a bottle shelf in the door surrounded in yellow interior completes the storage space in this 8.4 cubic foot model.

Cabinet dimensions are 29" wide, 29" deep, and 59 ½" tall.

M-85 and M-85L came with a defrost water container, defrost chill tray, crisper, three plastic grid ice cube trays and the Diffuse-O-Lite. Bottle openers remained standard.

M-85-D, M-85-DL
Featuring a magnetic bottle opener, two chrome plated steel shelves, twin crispers with glass covers, a defrost container with push button automatic defrosting, and large meat drawer, three pantry door trays and a in-door butter keeper, this 8.3 cubic foot refrigerator is the deluxe version of the previous model.
Cabinet dimensions are 29" wide, 29" deep, and 59 ½" tall.

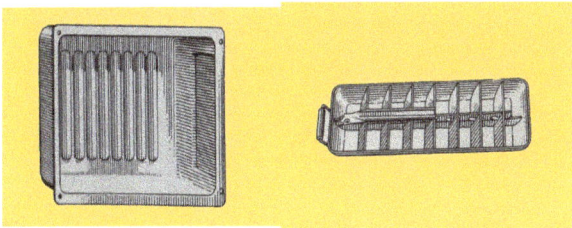

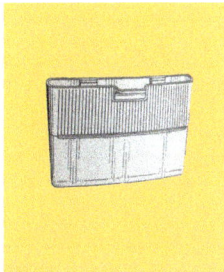

M-85-D and M-85-DL arrived with a one pound butter tray, two crispers, a meat drawer, defrost water container, and four lever type ice cube trays

M-104

The model designed to complement the M-series freezers, this unit features the U-shape freezer, with vertical bronze trim, large porcelain enamel steel meat drawer, three and a half chrome plated shelves, giant crisper with glass cover that doubles as a fifth shelf.

Temperature control, bottle opener in door strike and light remains standard on all refrigerators built from 1950 and onwards.

The big feature for this 10.4 cubic foot fridge is the in-door pantry. A heated butter keeper is surrounded by six shelves of food, made up of two egg racks, two bottle trays, and two utility trays that are held in place with bronze trim guards.

The exterior cabinet dimensions are the same as the M-85 models.

Cabinet dimensions are 29" wide, 29" deep, and 59 ½" tall.

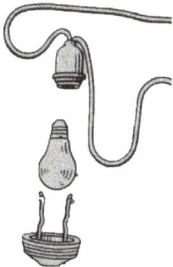

M-104 has four plastic grid ice cube trays, a meat drawer, full width crisper, meat pan cover, and a one pound butter tray.

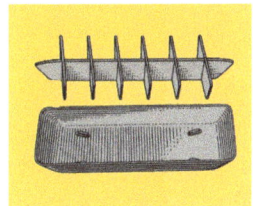

M-105-D

An automatic all-weather temperature control that feeds in cold air to replace warm air when the door is opened features in this deluxe model. Stainless steel shelves, push button defrost with water container, large crisper and meat drawer, four 16-cube ice cube trays with yellow lifter and magnetic bottle opener add to the luxury design. Bronze trim accentuates the full width freezer and door trays, as well as frame the sliding glass doors to cover the Bacon-Cheese-Butter Keeper. This thermostatically controlled area keeps bacon, cheese and butter at the ready for the perfect breakfast. This 10.2 cubic foot model with 17.4 square feet of shelf space features the iconic yellow of all the previous M-series models.

Cabinet dimensions are 29" wide, 29" deep, and 59 ½" tall.

Model M-105-D Household Refrigerator

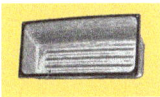

M-105 arrived with four yellow handle lever action ice cube trays, a full width crisper, meat drawer, defrost tray and defrost container.

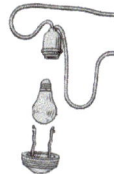

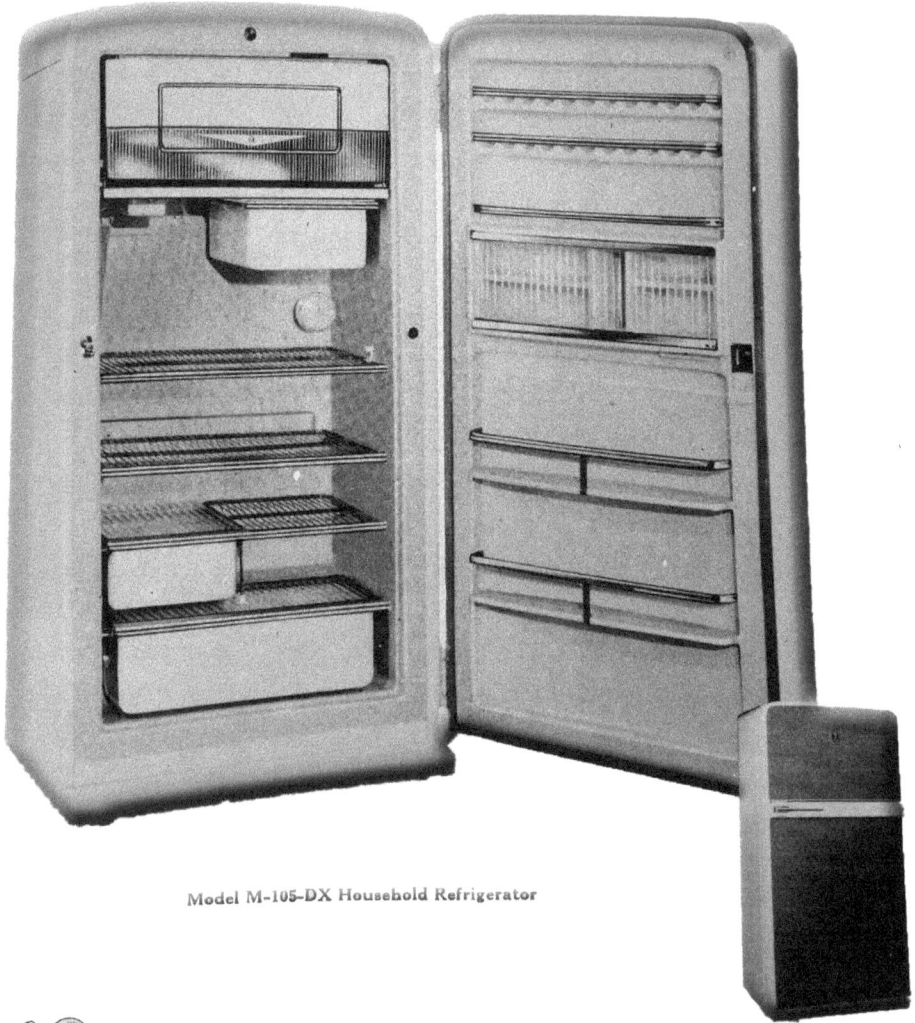

Model M-105-DX Household Refrigerator

M-105-DX

The M-105-DX offered all the features listed as well as a glide out shelf that rolled on nylon rollers, bronze ice cube trays with yellow lifters, and an additional half size crisper, making a total of 23 quarts of fruit and vegetable storage. Unit was shipped from factory with vinyl covering in a choice of colors. Terra-cotta, Spruce Green, or Citron Yellow were the choices, for upper or lower.
 Cabinet dimensions are 29" wide, 29" deep, and 59 ½" tall.

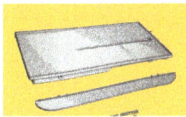

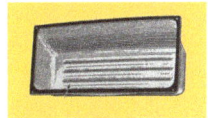

The M105-DX arrived with a one pound butter tray, one and a half crispers, meat drawer and cover, defrost shutters, defrost water container, defrost tray, and four lever style bronze ice cube trays with yellow handles.

Take your choice of the new **IH** freezers!

INTERNATIONAL HARVESTER

NO MATTER WHICH KIND YOU CHOOSE...

The flavor you **put in** is the flavor you **take out**!

New **IH** upright freezers

- 514 or 665 pounds capacity
- Takes no more space than a refrigerator
- Fastest defrosting of any home freezer
- Extra shelves and fruit chutes on the Super Pantry-Dor
- The only freezer you can decorate

There's no "frozen taste" in food you take from an IH freezer! Flavor can't change because food temperature never varies more than a degree or two—any place in the freezer, any time of the year.

Preset, automatically controlled temperature . . . Laminated glass fiber insulation . . . Exclusive "Dri-Wall" construction . . . Strongest freezer door in the world . . . Quiet, fanless, trouble-free operation.

See your IH dealer now. See how easy it is to own a new IH freezer!

New **IH** chest freezers

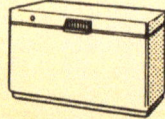

- Four models – 245 to 700 pounds
- Lid forms an extra work surface—which you can cover with counter-top material
- Extra flexibility with removable baskets
- More food space within easy reach

INTERNATIONAL HARVESTER   Freezers, Refrigerators, Air Conditioners

Advertising for 1955 freezers, upright and chest. TL

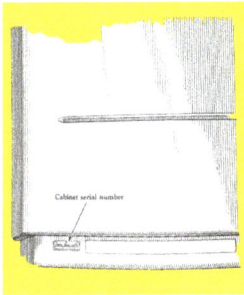

Serial Number Tags:
Have been moved to the front
left of all chest freezers on the
toe relief.

Serial Tags for upright freezers
are below the bottom hinge on
side of unit.

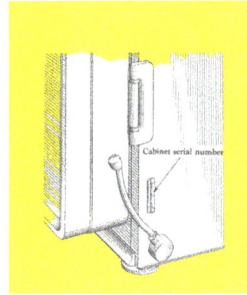

1955 A Series Freezer
A-7, A-12, A-16, A-20
AV-15, AV-19 Vertical Freezer

Introducing space saving freezers, yet keeping the same capacities, the convenience was appreciated with narrower and easier to reach interiors. Serial numbers were moved to the lower left front, on the toe recess of the chest freezers. On the vertical freezers, the small serial number plate can be found below the bottom hinge. Featuring sunshine yellow and copper trim to complement the refrigerator line, these freezers were the last of the units to be built in Evansville by International Harvester.

Trials of different sized motors for real use were common for Harvester. A thousand of each model, A-20 and AV-19, were built with ¼ hp motor instead of the standard 1/3 hp. These were identified by a stamp above the warranty label and on top of the compressor, RD-716 for the A-20 and RD-717 for AV-19. The serial number range for these are A-20 Serial #20151 through 21150 and AV-19 Serial #7501 through 8500.

Other freezers that also received the ¼ hp motor were; AV-15, starting with serial #12792; FPV-15 #501; FPV-19 #1286; A-16 #15396.

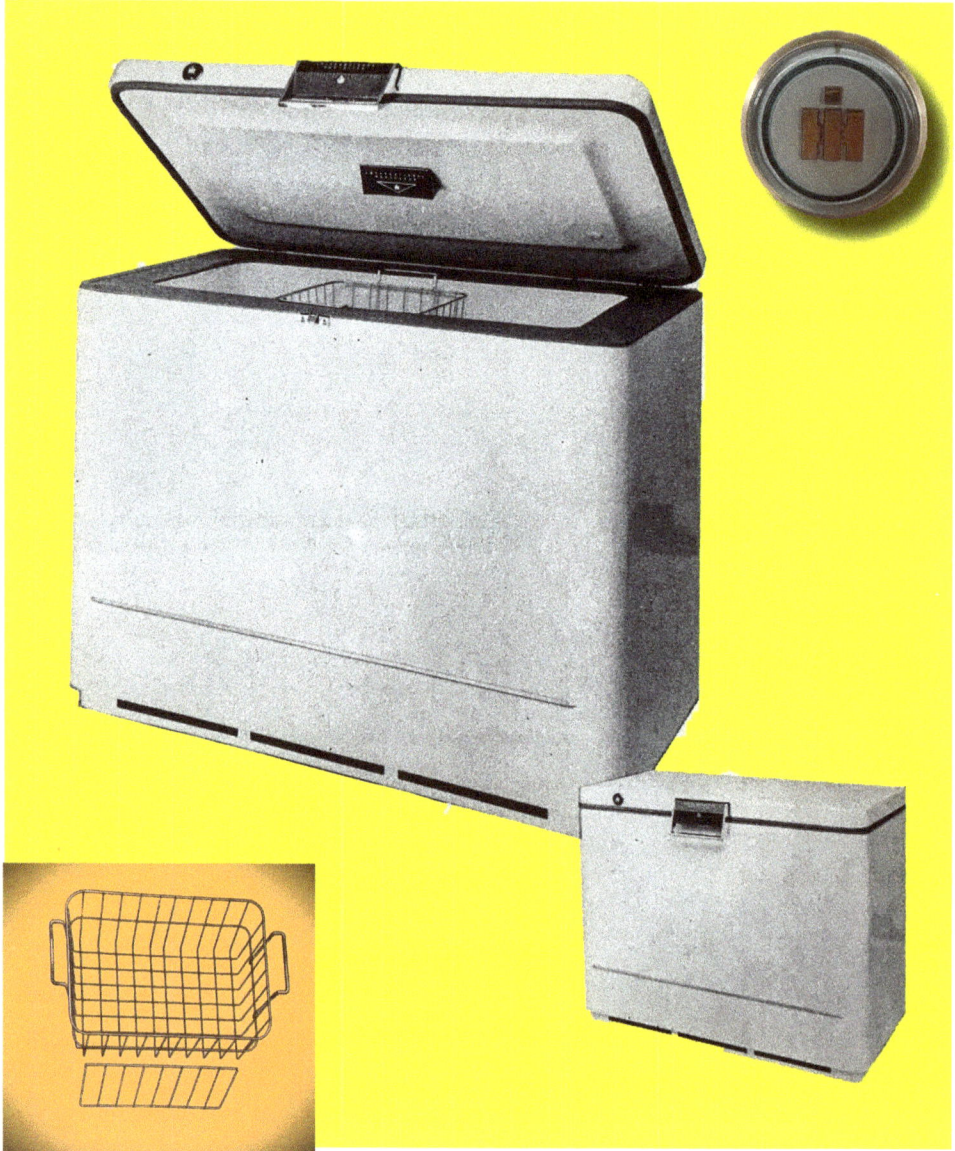

A-7, A-12, A-16, A-20

A-7
The small 7 cubic foot A-7 chest freezer arrived with sunshine yellow interior, copper accents, dome light, Lighted lens indicator, one basket and one divider remained standard, crated weight 325 pounds, and a list price of $269.95. Optional field attachment lock available for extra cost.

A-12

This 12 cubic freezer came with lockable handle, interior dome light, warning light, two baskets and two separators, one small and one large. Powered by a 1/4 horsepower compressor, arrival weight of 419 crated pounds, retail $369.95. Copper and Sunshine yellow theme throughout.

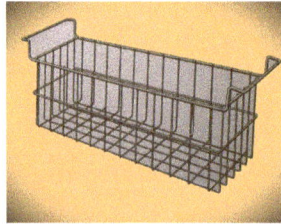

A-16
Three baskets, two large separators and one small arrived with this 16 cubic foot freezer, complete with dome light and lens, exterior operating light, decor option cover, locking handle and powered by a 1/3 horsepower motor.
$469.95 suggested price.

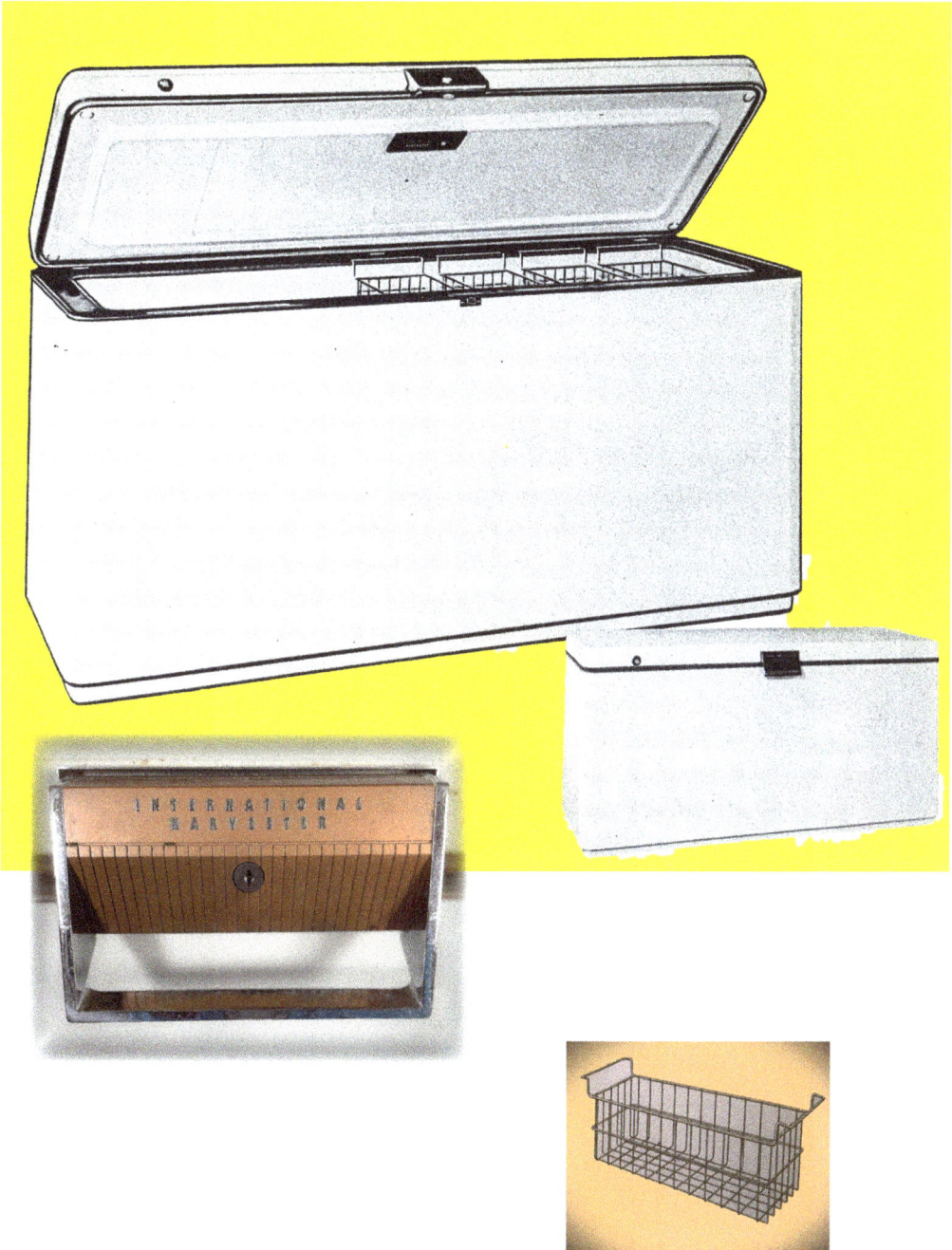

A-20
Crated weight of 579, four baskets, two large wire separators and one small wire separator, suggested price of $529.95. Lockable handle, warning operating light, interior dome light with lens, sunshine yellow interior, copper trim accents, brown rubber seal.
1/3 horsepower compressor.

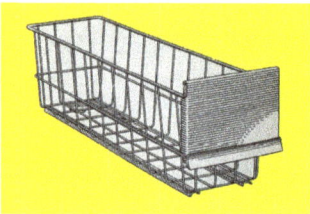

AV-15

Three refrigerated shelves, one adjustable, three in-door pantry trays, three rolling baskets. Interior light, one door mounted four can wide juice dispenser, all trimmed in copper. Crated weight 452 pounds, powered by 1/3 horsepower compressor. Suggested list price of $499.95

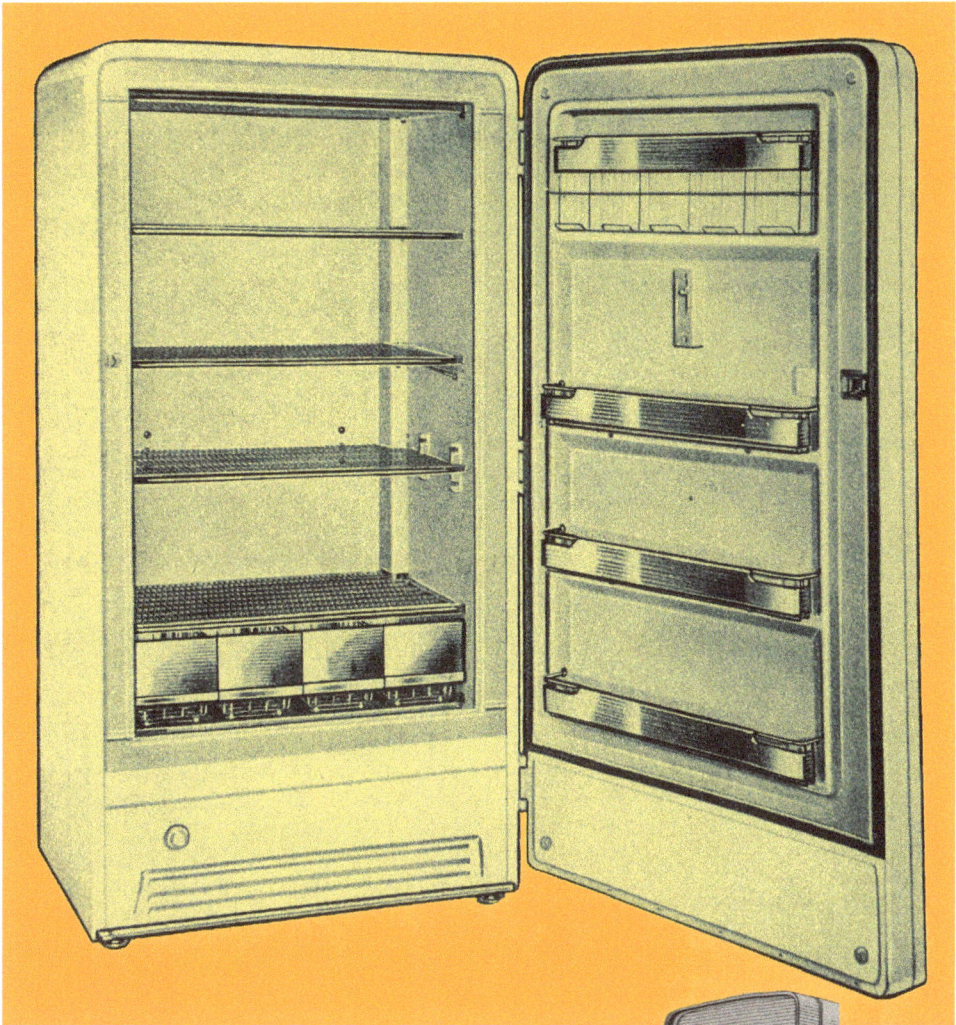

AV-19

Three refrigerated shelves, one adjustable, three in-door pantry trays, four rolling baskets. One door mounted five can wide juice dispenser, interior light, all trimmed in copper with sunshine yellow interior. List price of $579.95 arriving crated weight of 540 pounds, powered by 1/3 horsepower compressor.

A-Series Owners Manual, above. Top right, advertising for the new A-Series line.
Photo of the new refrigerators for sale in November, 1955 at Chesaning Sales Co,
Michigan. TC

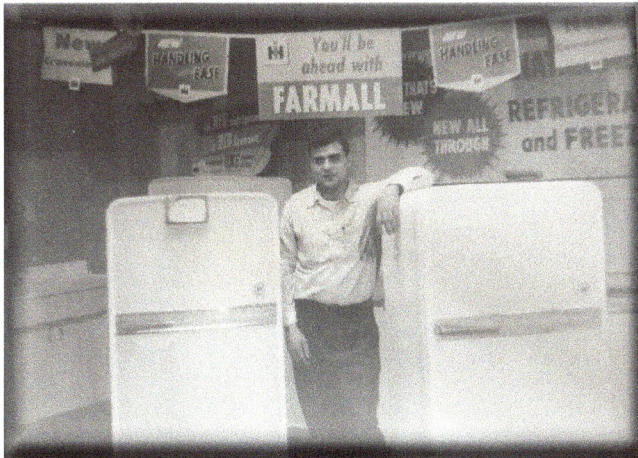

1955 A-Series Refrigerators

A-75; A-85; A-95-D; A-104; A-106-DX; A-106- D; A-120-D

Harvester continues to start with serial number 501 for all refrigeration units, choosing to move the serial tag location on this series to the side of the cabinet, below the bottom door hinge.

The last of the models before Harvester sold the Refrigeration Division to Whirlpool-Seeger, these models were the most advanced line.

Features like the Automatic defrost; Automatic All-Weather Control; Super Pantry Door; Adjustable shelf with glide-out shelves; "Tight-Wad" compressor; Porcelain Enamel interiors and many other features including the exterior Decorator Door were exclusive to the Harvester line.

Colors were Sunshine yellow with Copper trim, Brown rubber seals and gaskets.

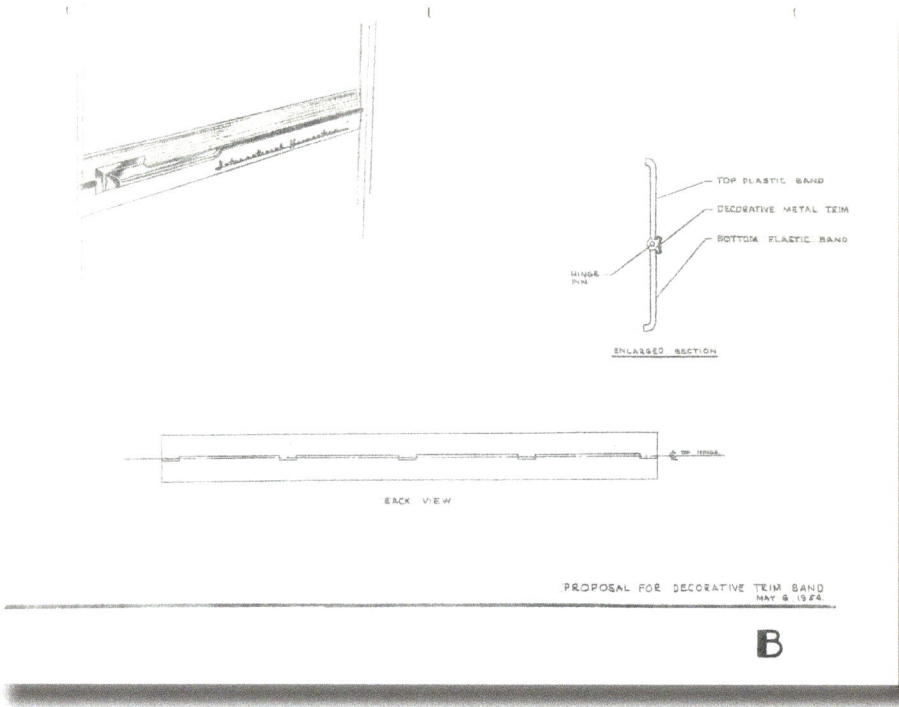

TOP PLASTIC BAND
DECORATIVE METAL TRIM
BOTTOM PLASTIC BAND
HINGE PIN

ENLARGED SECTION

BACK VIEW

PROPOSAL FOR DECORATIVE TRIM BAND
MAY 6 1954

B

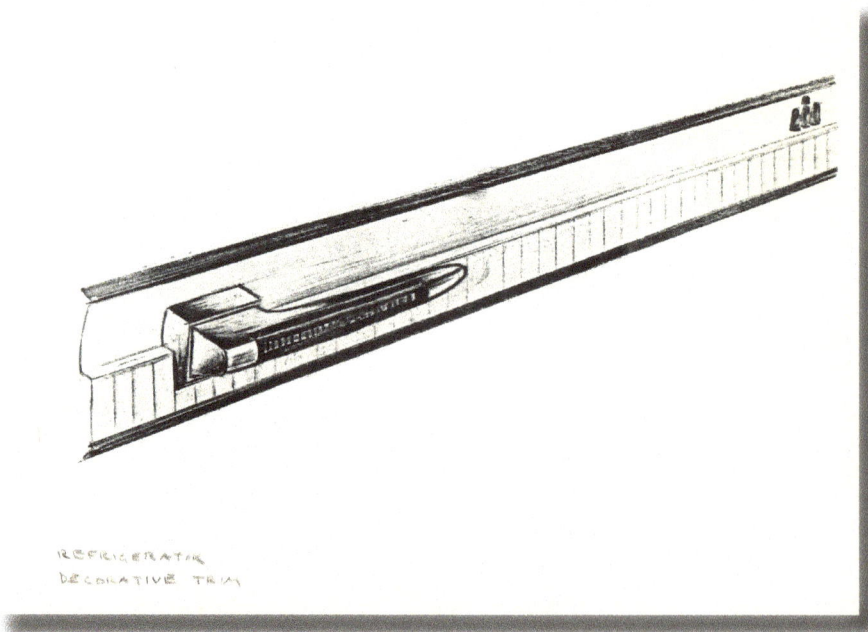

REFRIGERATOR
DECORATIVE TRIM

Two concept drawings for refrigerator handles, drawn in 1954 for the 1955 and 1956 design.
Development for new lines started about two years before planned release.
These two handle drawings show elements that were applied to the 1955 line, keeping the
continuity of the Harvester refrigeration image.
Drawings supplied by Greg Montgomery, former International Harvester Design Manager.

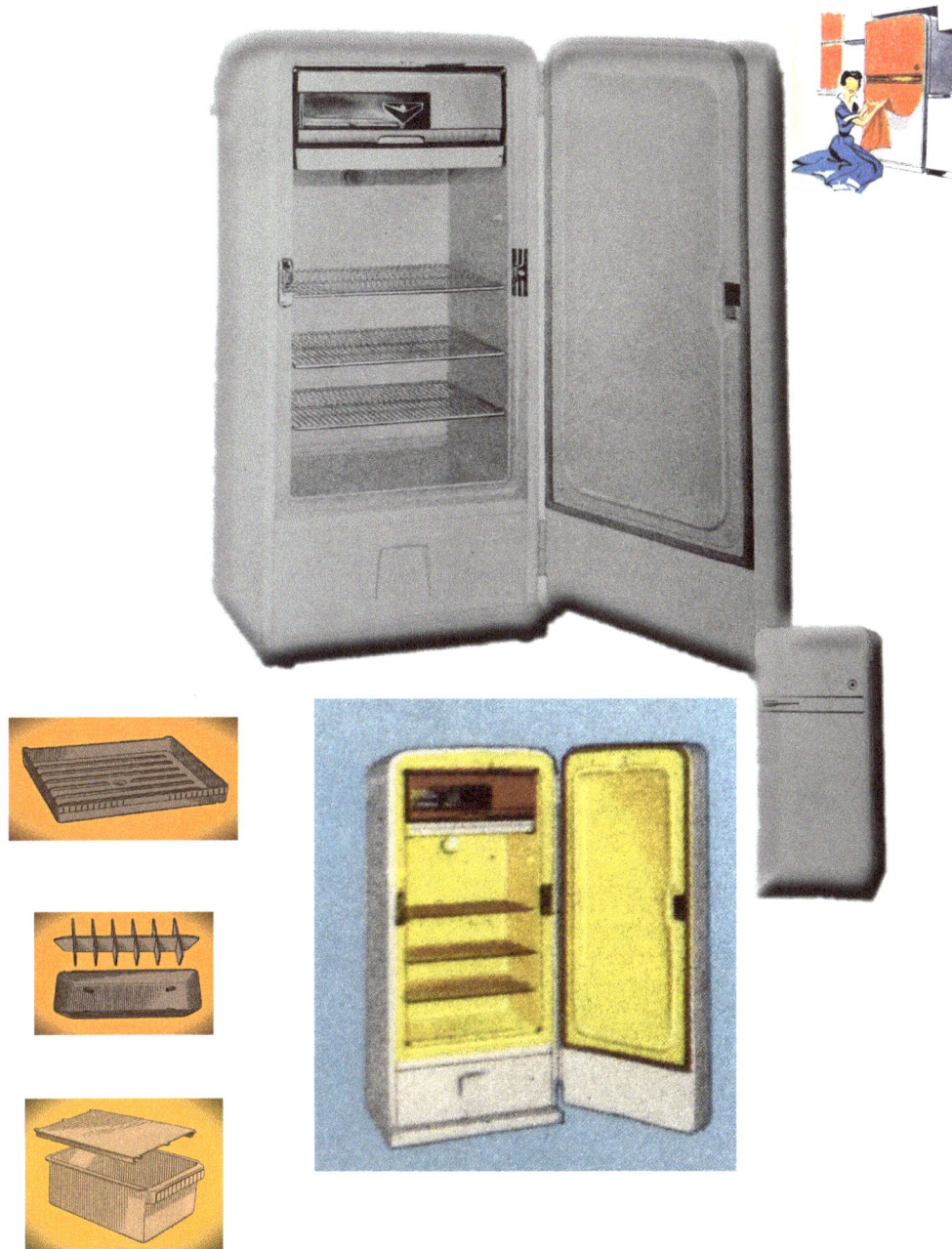

A-75, A-75L

Continuing the entry level unit for small spaces, 7.4 cubic feet features three zinc plated wire shelves, the popular clear view lift-up freezer door in copper coloring, two plastic grid ice cube trays and plastic defrost tray. Footprint for this unit is 25" wide, 28" deep, 54 3/8" tall. Powered by 1/8 horsepower compressor, this unit is shown in right hand opening, a left hand opening door was also an option. Both units were priced at $189.95. ($2,132.82 in February 2023) crated weight of 280 pounds. An additional attachment, ordered separately, was a porcelain crisper pan with cover.

Model A 85

Left-Hand Opening A-85D is still keeping drinks cold at the Agricultural Museum, U.P. State Fairgrounds, Escanaba, Michigan. This is operated by the U.P Steam and Gas Engine Association.
This refrigerator still has the original 1/4 pound Butter tray, and defrost water container. TC

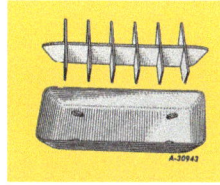

The A-85D and A-85 DL were introduced in September of 1955. These were similar to the A-85 Units, include the addition of a push button defrost.
The base mounted compressor was the same for A-85-D Serial 501 to 8011 and A-85-DL Serial 501 to 834.

Standard accessories were one glass quarter pound butter dish, two plastic grid ice cube trays, full width crisper, defrost tray under freezer compartment for all four units.

Defrost option introduced for the A-85D and A-85DL in September of 1955.

Left hand opening door indicated by 'L' prefix on models.

A-85, A-85L, A-85D, A-85DL

The introduction of this A-85 model in December 1954 did not include the A-85D or A-85DL. These two models were added to the series in September of 1955. The addition of a push button defrost, and base mounted compressor with related electrical components indicate these were designed for the 1956 series release. Serial number breaks for the changes were important when replacing the base mounted compressor for A-85D, as the new base began with serial number 8011 and for A-85DL, serial number 834. Earlier numbers of this model were to order as listed for A-85 or A-85L.

The A-85 model features the same outside dimensions as the A-75 cabinet. Inside features included three chrome plated shelves, a full width crisper drawer with lid, an insulated freezer door, with door becoming a shelf. A full width chill tray and two plastic grid ice cube trays complete the evaporator area. The inner door liner in yellow showcases two utility trays, two egg trays and a butter keeper trimmed with copper anodized rails. 8.5 cubic feet of storage includes a ¼ pound butter tray, all powered by a 1/8 horsepower motor. Retail price was $239.95 ($2,693.53 Feb 2023) Crated weight was 295 pounds.

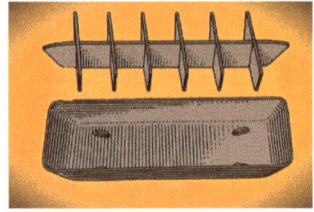

Three plastic grid ice cube trays, a quarter pound glass butter tray, water defrost container, defrost tray, crisper drawer, meat drawer,and egg storage all were found on the A-95-D refrigerator.

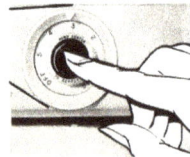

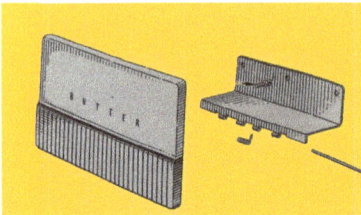

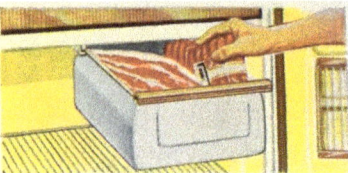

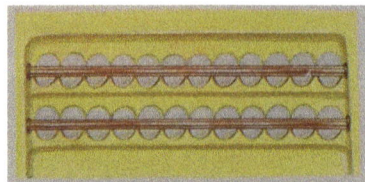

A-35498

A-95-D

Starting the lineup in the large cabinet size is the 9.4 cubic foot model with push button defrosting and large crisper. A freezer holding 50 pounds of food, three chrome plated shelves, four pantry door shelves (two egg, two bottle), three plastic grid ice cube trays, a butter keeper with ¼ pound butter tray, meat drawer with cover and defrost tray with water container rounds out the features.

Dimensions of the unit are 29 ½" wide, 28" deep, 59 ½" tall, powered with a 1/8 horsepower motor, crated weight was 366 pounds, and suggested list price is $339.95 ($3,816.63 February 2023)

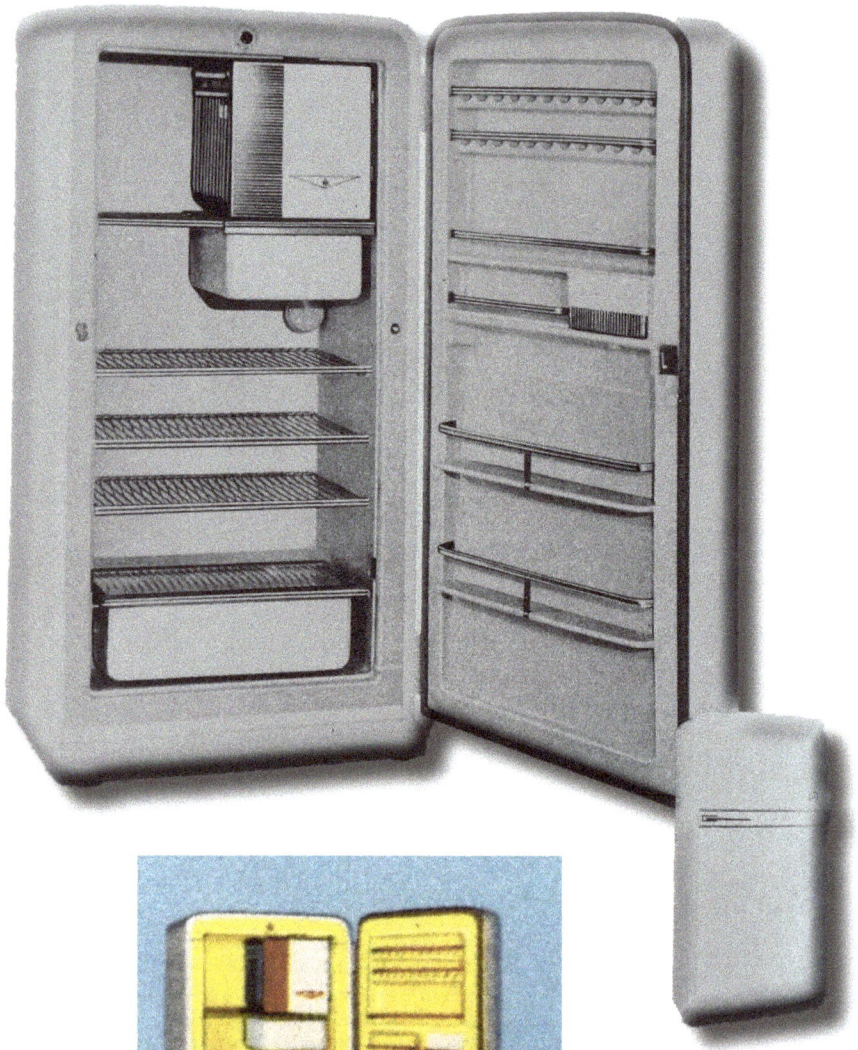

The A-104 arrived with full width crisper, meat drawer, three plastic grid ice cube trays, and a one pound butter tray.

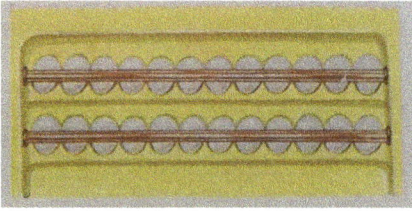

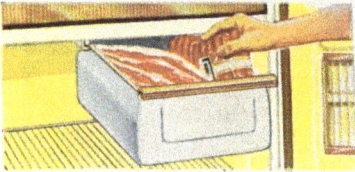

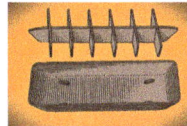

A-104

The large capacity refrigerator with U-shaped freezer is able to hold tall bottles with ease in this 10.6 cubic foot unit. A big meat drawer with cover, full width crisper and three plastic grid ice trays and three chrome plated shelves are complemented with the yellow interior and copper trim. The pantry door features two egg trays, two bottle trays, two utility trays and a heated one-pound butter keeper with tray. Crated weight of 361 pounds and recommended retail price $309.95. Dimensions of the unit are 29 ½" wide, 28" deep, 59 ½" tall, powered with a 1/8 horsepower motor.

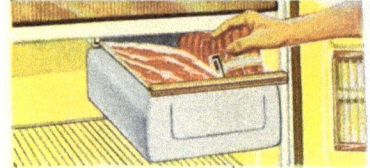

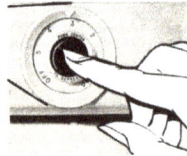

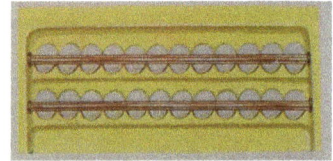

A-106D and A-106DX came with three lever release ice cube trays, one and a half crisper drawers, meat drawer, defrost water container, roll out shelf, a one pound butter tray and a bacon keeper.

A-106-DX, A-106-D

A deluxe refrigerator, all features are the same for both models, except for the A-106-D not supplied with vinyl door covers.

These door covers were regularly supplied for the DX model, and were optional on A-95, A-104, A-106-D. Colors available were Spruce, pink, primrose, charcoal, aqua, gray, persimmon for both upper and lower door panels.

Push button defrost, automatic temperature control, large fifty pound capacity freezer, three copper ice cube trays with yellow handle, defrost tray with water container, stainless steel shelves, one pullout, one standard, one half glass and wire. Two crispers, one full width and one-half width. Pantry door features seven shelves, including a heated butter, bacon, and cheese keeper.

Includes a sliding glass door cover, one pound butter tray, bacon tray and magnetic bottle opener.

Dimensions of the unit are 29 ½" wide, 28" deep, 59 ½" tall, powered with a 1/8 horsepower motor, crated weight of 388 and 379, respectively.

Retail Price of $424.95 for 106-DX and 106-D of $419.95.

A-35500

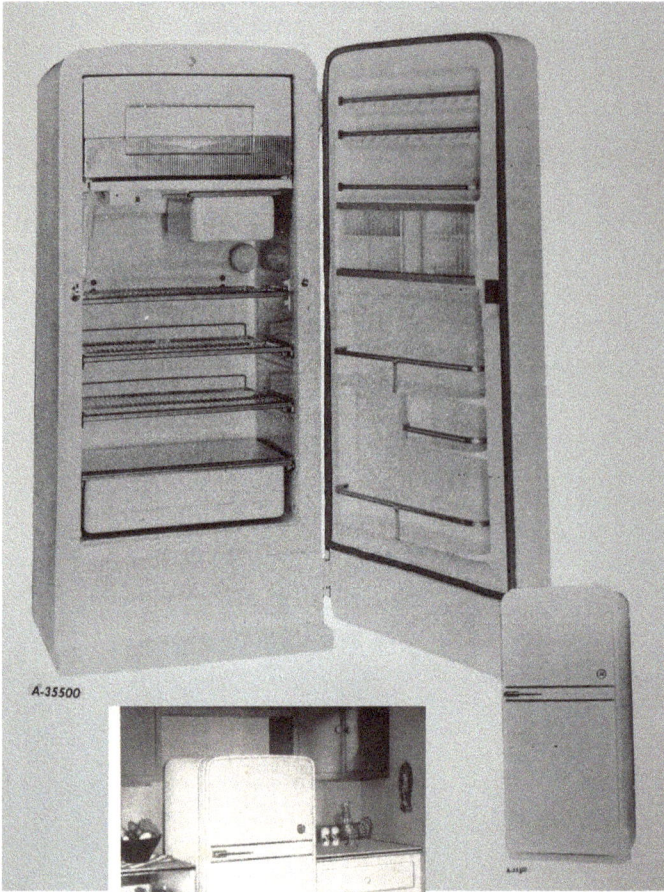

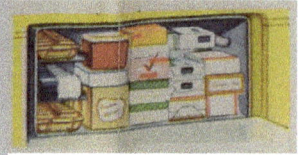

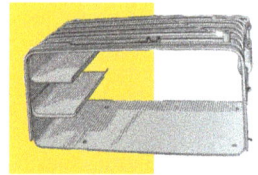

Standard accessories for A-120 and A-120 DL are Bacon tray, one pound butter tray, full crisper drawer, meat drawer, three lever type ice cube trays, defrost container, and two roll out shelves.

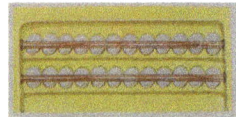

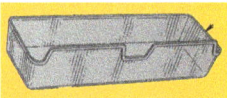

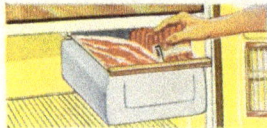

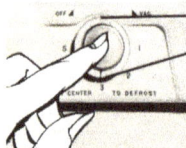

A-120-D, A-120-DL

A fantastic refrigerator for 1955, this unit stands four inches taller than previous units. Packed into 12 cubic feet of food storage is a 65 pound frozen food storage, with interior shelf for three copper colored ice trays with yellow handles and push button defrost.

Automatic all-weather control is mounted next to the meat drawer, and under the defrost tray. Three stainless steel shelves, two roll out, one adjustable, and a large, 29.4 quart crisper drawer with glass cover completes the shelf spaces.

The pantry door features two egg trays, a utility tray, two and a half bottle trays, and a heated butter, bacon and cheese keeper. Butter tray holds one pound of butter and the bacon container makes it easy to peel strips of bacon for a fresh sandwich. Magnetic bottle opener in door latch completes the 29 ½' wide, 28" deep, 63" tall refrigerator.

Door cover color choices were spruce, pink, primrose, charcoal, persimmon, aqua, or gray with a 3/8" wide, seven yard roll of double sided tape.

Crated weight of 387 pounds, Powered by 1/8 horsepower compressor.

Suggested List price of $449.95 ($5,050.86 February 2023)

Three women in the home economics model kitchen at International Harvester's Evansville Works. Zelma Purchase is at the table, Loris Knoll is at a stove, and Mrs. Ethel Jean Mitchell is at a sink. An International Harvester Model 8H3 and a Model 4 FC International Freezer can be seen. WHS 20726

"...it's femineered"

A very brightly colored test kitchen, with yellow walls, blue trimwork, red counters, gray cabinets and green accents welcomed anyone walking into the Test Kitchen at the Evansville refrigeration engineering and test facility in 1947.

For Loris Knoll, Priscilla Cobb, Zelma Purchase, Ethel Mitchell, this was a laboratory. The ladies spent their day testing the products for best refrigeration, freezing, and storing techniques. Led by Ruth Whiting, Director of the home economics department, the kitchen developed uses for products; presented the technical advances made to dealerships; and trained countless home demonstrators throughout the country that presented to communities how to store or freeze foods.

This kitchen also tested recipes, researched products and ate copious amounts of good food of many varieties, before and after storing in the refrigerator or freezer. This approach by an all female staff to the engineering side of food storage led to the slogan "...it's femineered!" The five ladies also were the elements behind the fictional character, Irma Harding. Irma was the spokesperson for all packaging, storage containers and accessories recommended by these ladies. Irma also was the face for the recipe books and helpful hints that were a part of buying these appliances.

This approach to food in the home was not new to Harvester, as the Service Bureau (1912-1932) had presented and printed thousands of booklets for growing gardens, food preservation via canning, tinning and dehydrating and more to communities as a way to help people and towns grow, prosper and succeed.

Air Conditioners came in a variety of options to suit the room size. The larger 1 hp units were tightly fitted into the same cabinet as the small 1/4 hp units. TC

Dehumidifiers and Air Conditioners

International Harvester 's Refrigeration division offered refrigerators and freezers and they also sold in the last three years of production, dehumidifiers, and air conditioners. If the company could make the air cold in a small area it only makes sense that they can make our everyday living area cold. The window air conditioners were offered in a variety of finishes, or you could also add your own fabric. Yes, Harvester sold fabric!

Air Conditioners

The first introduction to the general public of mechanical air refrigeration occurred at the St. Louis World's Fair in 1904, in the Missouri State Building. Designed by Carrier in 1902, to solve a humidity issue in the Sackett-Wilhelm Publishing company to prevent magazine papers from wrinkling, the method of drying air is a comfort in many homes today. Willis Carrier was employed by Buffalo Forge, a large company that may have supplied the lever driven fan style forge to International Harvester's Foundries or part of the air handling equipment in the Factories. Buffalo Forge fans still heat or cool some of the most iconic buildings in the world.

In the beginning of this technology, cooling air was cumbersome and expensive and not remarkably effective. Movie theaters that utilized air conditioning would be freezing cold at floor level and stifling hot on upper levels. In 1922, Carrier designed a system that would help mix the air more effectively. This design also used fewer moving parts and was more dependable. Window units were invented in 1932 and by 1947 were more affordable to most houses in America without major changes or upgrades to the home.

Refrigeration of air was considered a health choice, and doctors recommended installing a unit for the benefit of those that suffered from heat. Ideal units were those that expelled the air upwards, and did not create direct drafts, causing chills. An elderly person could also suffer from chills if the unit did not have a thermostat that shut off the compressor when the ideal room temperature was reached. The fan needed to stay on to circulate the air in the room, helping to prevent sudden changes in degrees. Ideal air conditioners needed to cool, clean, dry, and filter the air for increased summertime energy and appetites, preventing irritability that hot muggy weather causes. If you were living in the 1950s, wouldn't you want to buy an air conditioner after reading about those health benefits?

Harvester offered air conditioners for sale in the 1953 L-Series refrigeration line of products and continued until 1955 with the close of the refrigeration line.

Sales points included no chilly drafts, more cooling capacity, filtered air and the quietest operation compared to the competition. If you have ever experienced one of these running, you wonder just how loud the others were. If you have not had this luxury, be thankful you still have your hearing. As part of the Decorator Series, the units were color changeable in the front to match the room in which it was installed. The air blew upward, and could be directed left or right, forward or backward, another advantage over the competition's models.

The lightest model weighed 115 pounds, and the heaviest 198 pounds.

All-seasons models featured a heat pump as well, which could reverse the cooling process and heat a room. These were available in the 1954 M-Series and 1955 A-Series.

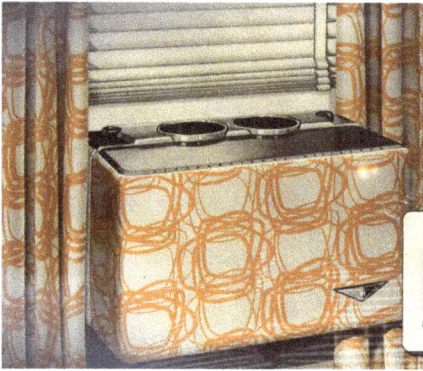

SEE THE NEW

Decorator
Air Conditioner

with the changeable fabric finish

THE SMART WAY TO KEEP COOL!

by International Harvester

Year	All L (1953), M (1954), and A (1955) Models						
Deluxe Models	**350-D**	**500-D**	**750-D**	**751-D**	**752-D**	**1000-D**	**1001-D**
"All Seasons" Models	--	**600**	**850**	**851**	**852**	**1100**	**1101**
Compressor	1/3	1/2	3/4	3/4	3/4	1	1
Power, A.C, Volts	115	115	115	230	208	230	230
BTU per Hour	4000	5500	8500	8500	8500	10600	10600
Moisture Removal (pints/hr)	1.3	1.75	2.5	2.5	2.5	3.25	3.25
Width in inches	22 3/8	26 1/4	26 1/4	26 1/4	26 1/4	26 1/4	26 1/4
Height in inches	12 1/4	15 5/8	15 5/8	15 5/8	15 5/8	15 5/8	15 5/8
Projects inside window	12 1/4	14 7/8	14 7/8	14 7/8	14 7/8	14 7/8	14 7/8
Projects outside window	13 3/4	16 3/8	16 3/8	16 3/8	16 3/8	16 3/8	16 3/8
Air delivery, CFM	140	240	270	270	270	300	300
Room Air exhaust	30	125	130	130	130	140	140
Fresh air	35	110	110	110	110	130	130
Thermostat	Yes	Yes	Yes	Yes	Yes	Yes	Yes
Net Weight, pounds	115	163	187	187	187	198	
Refrigerant Type	R-12	R-12	R-12	R-12	R-22	R-22	R-22
Original Amount, Ounces (+/- 0.5)	18.0	15.5	23.0	23.0	22.0	27.0	27.0
Replacement Motor, Ounces (+/- 0.5)	19.0	16.5	24.0	24.0	23.0	28.0	28.0
New Filter Drier on both							

Year	All AR Models (1956 Series)					
Models	**AR-40**	**AR-50**	**AR-70**	**AR-71-D**	**AR-100**	**AR-100-D**
Compressor	1/3	1/2	3/4	3/4	1	1
Power, A.C, Volts	115	115	115	230	230	230
BTU per Hour	4100	5500	6800	6800	11000	11000
Moisture Removal (pints/hr)	1.3	1.75	2.5	2.5	3.25	3.25
Width in inches	22 3/8	22 3/8	22 3/8	22 3/8	26 1/4	26 1/4
Height in inches	12 1/4	12 1/4	12 1/4	12 1/4	15 5/8	15 5/8
Projects inside window	12 1/4	12 1/4	12 1/4	12 1/4	14 7/8	14 7/8
Projects outside window	13 3/4	13 3/4	13 3/4	13 3/4	16 3/8	16 3/8
Air delivery, CFM	140	180	210	210	270	300
Room Air exhaust	30	30	30	30	140	300
Fresh air	35	40	50	50	130	130
Factory installed Thermostat	No	No	Yes	Yes	No	Yes
Net Weight, pounds	115	163	187	187	187	198
Refrigerant Type	R-12	R-12	R-22	R-12	R-22	R-22
Original Amount, Ounces (+/- 0.5)	18.0	19.5	19.0	19.0	27.0	27.0
Replacement Motor, Ounces (+/- 0.5)	19.0	20.5	20.0	20.0	28.0	28.0
New Filter Drier on both						

The only Air Conditioner you can decorate

Your IH Air Conditioner comes to you decorated in beautiful fabrics—in Mocha (Beige), Flame (Red), Olive (Green), or Mustard (Yellow). Without fabric, it's an attractive neutral sand finish.

Best of all, you can change the fabric front to match your draperies, your window trim or walls whenever you wish . . . in minutes . . . with less than a yard of fabric . . . in any color or pattern you choose.

THERE'S AN IH AIR CONDITIONER JUST RIGHT FOR YOU

See them on the following pages — then send this coupon for an estimate

Air Conditioners were sold for three years, from 1953-1955. 1956 Series were released in late 1955.

International Harvester... the Air Conditioner that gives you *true* **No-Draft Cooling**

Here is the air conditioner that's truly draft-free, truly safe — yet it cools more air, faster, at lower operating cost.

New IH Air Conditioners send cool air out at the *top* — direct it up to the ceiling so that it spreads out evenly and settles gently over the *whole* room— and *never* blows a chilling draft directly *at* you.

Cool air from an IH goes UP . . . Settles gently over the whole room . . .

Never blows cold air directly at you.

Heats on chilly days, too.

International Harvester Air Conditioners are designed to give you maximum comfort cooling without drafts.

R-20.

OWNER'S MANUAL

INTERNATIONAL HARVESTER DEHUMIDIFIER

INTERNATIONAL
HARVESTER

MODEL L-24

INTERNATIONAL HARVESTER
Dehumidifier
MODEL M-24

"Don't mess with a good thing" Harvester must have decided that the dehumidifier did not need improvements other than a on/off switch on their unit. A separate timer could be added, if desired. TC

THE SMART WAY to guard almost everything you own: from damaging air moisture. Prevents rust, mold, mildew and rot by removing excessive moisture from the air. Clothes, books, tools, rugs, luggage, and other valuables can be safely stored without fear of damage; restores utility rooms, basements, store rooms, recreation rooms to dry, pleasant, usable areas. If you have a moisture problem, you can end it with a new International Harvester Dehumidifier.

Dehumidifiers

The function of a dehumidifier is to remove excess moisture from the air to prevent mold and mildew, which can cause damage to clothes, books, tools or any room that is excessively humid that should be dry. Defined as a Condensate dehumidifier, these units use a refrigeration cycle to pull air across a cold surface, causing the water to condense on the desired surface. The dry warm air is circulated in the room, effectively reducing the moisture in the air. This dehumidification was part of the development by Carrier when tasked with keeping the presses dry and free from rust in 1902.

Introduced in 1953, it was designed to remove one pint per hour, or 24 pints in 24 hours from the air. The L-24 model removes water at 80% relative humidity in a room size up to 10,000 cu ft. 1/9th horsepower, 115 volt and was sold through 1955. M-24, built for 1954 sales year, these two models were the only dehumidifiers produced. The M-24 featured an on/off switch on the side, where the L-24 does not have a switch. These two units were the same size. The serial tag for the dehumidifier is located on the upper corner of the inside of the unit body. The condensing motor serial tag is separate and located on the base of the condensing mount. Three ways to collect water, by using a ten-quart pail, provided, or connecting a hose to the drain, setting the unit over a drain or a washtub. The unit weighs 52 pounds and measures 22 ¼" tall, 17 7/8" wide and 13 1/2" deep.

A service bulletin dated September 21, 1953, was sent to dealers informing them of an IH approved timer for their dehumidifier sales. This was shipped direct from the Badger Products Company in Neenah Wisconsin. This was to assist in the control of the dehumidifier, limiting the amount of run time.

A worker is performing an inspection of the M1 Garand produced at Evansville, Indiana. WHS 24154

Next page, the T-12 pilot unit for the US Military, without gun attachments. This APC has helped in military engagements to successfully keep troops safe. WHS 12133

Defense Production

International Harvester had always been a part of supplying military goods to the US Government, most notably in World War II, by converting nearly every inch of every factory over to war production. Harvester Factories produced a wide variety of items and continued to provide even in peacetime. With the start of the Korean War in 1950, various factories were building airplane wheel and brake assemblies, airplane gear landing struts, fuses, and other component parts. The majority of contracts that Harvester fulfilled consisted of vehicles, mainly the five-ton military truck used for a variety of purposes as well as the armored personnel carrier. Production also included M1 Garand rifles beginning in 1952.

One of the only times that the M1 Garand was shown to the civilian public during manufacture was at the 1953 Minnesota State Fair. The display included a complete breakdown of the components and parts. It was displayed alongside the T-18, an armored personnel carrier (now known as the M-75). This is a full tracked vehicle that played an important role in the evacuation of American soldiers under direct fire from Pork Chop hill near the end of the Korean War. The display was created in conjunction with the Army Ordnance Corps, to honor the Seventh Division soldiers who were withdrawn from Communist fire without a single casualty during the operation.

The Harvester company continued to produce the rifles for the Ordnance Corps, ending production shortly after the announcement of the sale of the property in 1955. Harvester negotiated termination of the M1 Garands to finish production by the end of March 31, 1956.

The total of this contract was about 337,000 rifles produced.

Sale of Company

International Harvester's decision to sell off the refrigeration in 1955 was a surprise to many. Record sales for July and August in nearly every refrigeration line showed a strong recovery from the previous 1954 report of nearly $52.8 million in sales to a close of $55.5 million in 1955.

On September 27, 1955, Harvester announced the end of the refrigeration division and sale to Whirlpool- Seeger Corporation. Harvester started the refrigeration division with the production of milk coolers, expanded into freezers, refrigerators, room-sized air conditioners and dehumidifiers. Quality and reliable products provided brought a reputation to the customer. With the rise in household refrigeration sales, naturally followed a consumer demand for other appliances. Rather than invest into a household appliance division, the company chose to sell a profitable line for $19 million to Whirlpool-Seeger.

The companies in the home appliance industry viewed Harvester as a short-line appliance company, the opposite of their well-known agricultural products.

The merger of the Whirlpool and Seeger and RCA corporations brought together the well-known products of Whirlpools home laundry products of washers, dryers, and irons, with the stove and air conditions of the RCA. The addition of the freezers and refrigerators from Harvester gave the Whirlpool Corporation a full line of home goods to offer to the metropolitan areas, a stronger presence than Harvester was able to achieve in the urban areas.

Whirlpool began company life in the laundry business, and in 1940, they were one of the twenty-six laundry machine manufacturers. By 1955, only four of the independent laundry companies remained. The Whirlpool company was in the same predicament that Harvester was. Either diversify or leave the household appliance business. Harvester chose to leave the business, as they were in the business of farming. Whirlpool chose to stay in the business and become the tenth company to offer a diversified full-line of home appliances.

The announcement of the sale and close of plant by Harvester to Whirlpool did not end Harvester's commitment to its 3,500 employees. Totaling nearly $2.5 million, severance checks of one week's pay for every year of service to each employee was one of the two steps to help the workers. The second was to make a concentrated effort to relocate employees elsewhere. Harvester was known to move employees as they advanced in their careers with the company.

The day after the announcement, representatives from twenty Harvester facilities were present for recruitment. The refrigeration division became a job recruitment center. The employees that did not want to leave the Evansville area were presented with list of prospective employers with rosters of more than one hundred job classifications from a six-man committee. Additional job contracts were presented as employment ended when the last refrigerator rolled off the line on December 7,1955.

Harvester had already found new employment or relocations for more than 2,500 workers. Whirlpool interviewed and hired many of them. Other out-of-town companies set up employment offices in the local hotels.

Several more companies set up a space inside the plant to interview candidates.

Two hundred of the 326 managerial staff were moved elsewhere with Harvester. The remaining 126 were placed in Evansville or nearby. Only eight were not relocated or hired elsewhere with Harvester's help.

Over the ten-year span of the Evansville Works, over two million refrigeration units were made. All of them were of high quality and long-lasting service. Both employer and employee had found a happy place in Evansville, and with the ten-year success of the plant, there was great sadness in this decision.

Although International Harvester's Refrigeration Division was short-lived in comparison to the rest of their products, they were influential and highly impacted on how the whole farm was more productive and profitable.

Today you can still find these refrigerators and freezers that were made from 1946 to 1955 in many garages, workshops, and basements. These freezers and refrigerators are heavy and large and awkward, somewhat collectible, and still in use.

The Whirlpool company continued to produce refrigerators at the Evansville plant, employing 6,500 people across three Evansville locations by 1969. In 1975, it closed the original Serval plant where gas refrigerators were built, due to a lack of demand. The Seeger-Sunbeam location was closed in 1984, and production of air conditioners, dehumidifiers and compressors was moved to LaVergne, Tennessee.

Production at the former International Harvester facility in the 5400 block of U.S 41 North ended in 2010 and the factory closed in 2014. In 2023, the buildings are used by at least four different businesses where Harvester once called home.

Whirlpool is still based in Benton Harbor, Michigan, where the company started in 1911 by Lou Upton. Some of the brands that are sold by them include Hotpoint, Amana, KitchenAid, and more.

Without the merger of Seeger, Sunbeam, RCA, and the purchase of International Harvester in 1955 by Whirlpool, where would our comforts in the home be today?

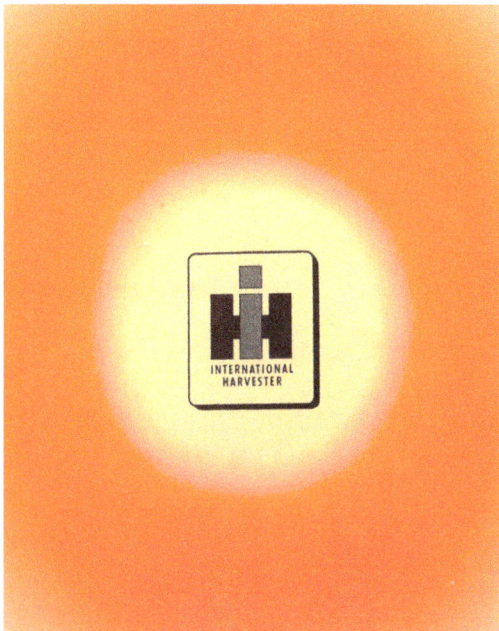

MODEL CA-71

MODEL CA-96

MODEL CA-96D

Three Canadian models, CA-71 (7.1 cu ft);
CA-96 (9.6 cu ft); and the CA-96D began sales in late 1954.

Your

International Harvester Refrigerator

Owner's Manual
and Warranty Record

INTERNATIONAL HARVESTER CO. OF CANADA
HAMILTON   ONTARIO   CANADA

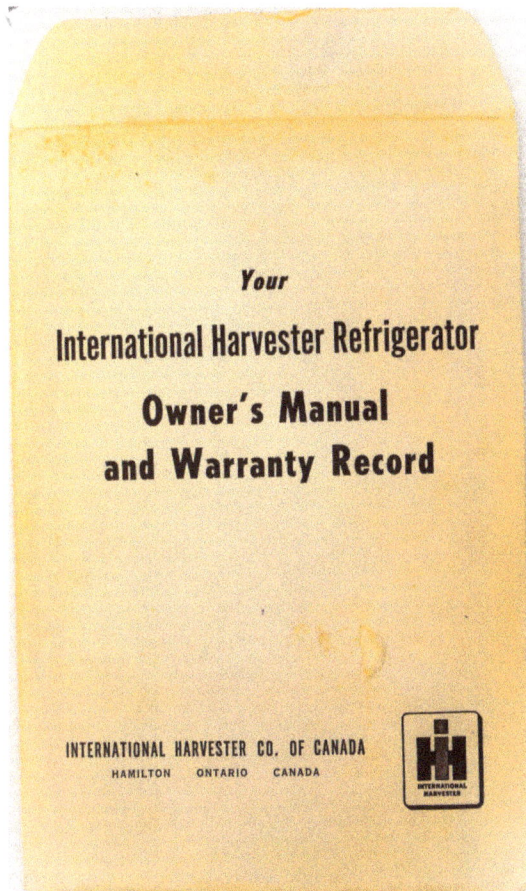

Overseas Build

Canada

The two other countries that made refrigerators for International Harvester were the Canadian and the Australian Divisions. The Canada market imported many from the United States, and in 1954 to keep up demand, the Canadian division signed a contract with Kelvinator of London, Ontario, Canada. Models produced beginning in 1954 were CA-71, CA-96, and CA-96-D.   Features on all three models were similar to the USA made version, with a butter dish and compartment or shelves in the door, crisper pans, meat chilling area and full width freezer. CA-71 was the entry level price model, with a single crisper drawer instead of two drawers on the next model. CA-96 and CA-96-D were similar in layout, with the addition of a cheese tray next to the butter tray, both located in the interior of the door. The Deluxe version, CA-96-D, featured roll-out shelves, and chrome plated trim in the interior.

When the closing of the Refrigeration division was announced in 1955, this division was also closed.

Advertising for 1956 in Canada shows the upcoming models; CB-81, CB-81-D, CB-105, CB-105-D which were never sold.

International Harvester REFRIGERATORS

featuring the MODERN TREND . . . two-tone beige interiors with coppertone trim

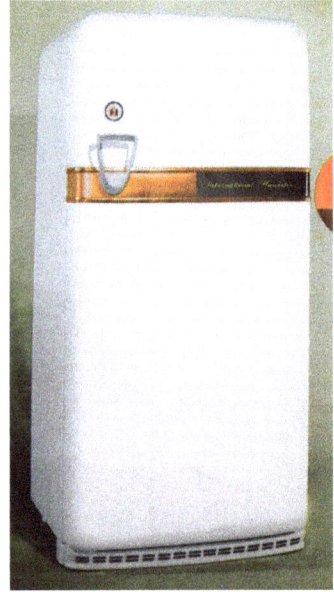

Modern Beauty

WITH ADVANCED STYLING

Beautifully styled, chrome plated door handle, which attractively sets off the sparkling beauty of International Harvester Refrigerators for 1956.

Coppertone and black panel reflects the latest decorator design trends; and makes your new refrigerator the featured showpiece of your kitchen.

This symbol, is your guarantee of dependability, long life and satisfaction in a refrigerator. The IH symbol represents an organization that builds a full line of motor trucks, industrial power and farm equipment; with a network of district offices and dealers from coast to coast throughout Canada to serve you.

INTERNATIONAL HARVESTER

1956 International Harvester "Extra Value" Refrigerators are priced surprisingly low

SEE THEM SOON!

These images from a Canada sales flyer show the new look for 1956. Slight differences for the new line include a cheese tray in addition to the butter tray, new handle styling, and for the top of the line model, roll out shelves.

CB-81, CB-81-D, CB-105, CB-105-D were the four models prepared to be sold.

Images from the Loschen Collection.

New 1956 "Space-Saver"
International Harvester
HOME FREEZERS

Planned for Better Living

Australia

In Australia, the refrigerators were made by Electrolux Australia for International Harvester Australia, and they were named Defenders. Two types of refrigerators were built, a kerosene or electric version. The electric was not available until 1953. Grid power was not common nor reliable in Australia, and in some places, stations (farms) still run on a generator in 2023. Kerosene was much more reliable when the power was not. Using just under a gallon of lighting kerosene, it was a very affordable option every week. With the introduction of the new electric model, names were added to define which unit was being referred to. Kerosene was called the "Country Style" version, and one operating as electric, referred to as "Selectra Style".

Ad in Australian Womens Weekly, October 5, 1946, for a kerosene refrigerator, two years before the models in USA were introduced. Tomac Collection.

There were three kerosene versions available from 1946 to 1956. The first unit was advertised in "The Australian Women's Weekly" October 1946, a popular women's magazine, two years before refrigerator production in Evansville. This first refrigerator, a kerosene only unit, was a 5 ½ cubic feet, 80-cubes in 4 trays ice maker, unchanged and available in 1954 for £121/5/- or about $242.50. In 2022, this would be equal to $4,749.27 Australian Dollars.

Advertising for the 1946 to 1952 kerosene refrigerator, left.

1953 to 1955 Kerosene refrigerator, right.

Below: Advertising in 1954.

Below Right: Kerosene refrigerator in need of restoration. All TC.

The 1953 Country Style was 7.1 cubic foot in size, with a full width freezer, and additional space for 80 cubes in five ice cube trays. It featured an insulated plastic meat chiller, thirteen square feet of zinc plated shelf space, and three storage drawers. Two smaller drawers were meant for eggs in one and fruits in the other. The large middle drawer also had a cover, for use as a vegetable crisper. This was advertised in 1954 for £159/10/- or $319. This would convert to $6,247.49 Australian dollars in 2022.

1953 advertising for the new style kerosene refrigerator. TC

The third model ended before it could be sold. When Harvester announced the closing of the Refrigeration division, worldwide was affected. This model release was prepared and advertising for January 1956 shows it was 7.3 cubic feet. It had a full width freezer, insulated meat tray, and two clear front crisper drawers. One of the two major changes for this model were in the ice cube making of three aluminum trays and one plastic for a total of 60 cubes. The other improvement was the addition of .2 cubic feet with three in-door shelves for eggs, small packages, and drink bottles.

Start the New Year RIGHT with a

1956 DEFENDER de Luxe

FAMILY SIZE 7.3 CU. FT. KEROSENE OPERATED REFRIGERATOR

MORE specialized storage space

FULL-WIDTH FREEZER LOCKER
Clear across the top the freezer locker holds food below freezing temperature.

SHELVES IN DOOR
Three shelves provide front-of-the-shelf storage for eggs, cans, bottles and small packages.

60 ICE CUBES
Three aluminium and one plastic tray make 60 ice cubes.

INSULATED PLASTIC MEAT CHILLER
Double base prevents meat freezing to plastic surface.

TWIN PLASTIC CRISPERS
Clear plastic crispers with "crystal-like" front are easy to keep clean.

LOOK at this wonderful new kerosene operated refrigerator! Look at these "specialized storage" features including shelves-in-the-door and twin plastic crispers. There is a full-width freezer locker, insulated plastic meat chiller, over 13 sq. ft. of shelf area and a glistening white cabinet to add glamour to your kitchen. Get full information on the Defender de luxe refrigerator from your local International Harvester dealer now!

INTERNATIONAL HARVESTER COMPANY OF AUSTRALIA PTY. LTD.

INTERNATIONAL HARVESTER

District Sales Offices in Capital Cities. Works: Dandenong and Geelong, Victoria.

AD. No. GL635-56 (8 inches x 10½ inches) — B. & W.
International Harvester Company of Australia Pty. Ltd.
For appearance in Country Weekly Newspapers commencing January, 1956.
Prepared by Advertising Dept., Merchandising Service, Melbourne.

This was advertising copy that was sent to the dealers for them to send to their local papers for printing. Dealers would insert their name and address in the corner when submitting for print. TC

The electric model first appeared in 1953, referred to as Selectra Style. The difference in the 7.1 cubic foot unit was the meat tray was full width below the freezer, same amount of shelf space and drawers. Advertised as "Double Duty" because it refrigerated as well as froze food items. The full width freezer held 35 pounds of frozen food, like ice cream and ice cubes. This model featured an additional half-width shelf. In 1954, £189/10/- or $379 plus freight to Longreach, Queensland would purchase this refrigerator. That would be equal to $7,422.57 in 2022.

1956 pre-advertising shows the electric refrigerator having an impressive 9.36 cubic feet interior size. The meat tray, full-width freezer and shelving remain the same options as the previous model. The updates for the model were an in-door butter keeper, three in-door storage shelves, one for eggs and two for tall drink bottles and frequently used items. This model was ended before production began for the 1956 sales year with the closing of the IH Refrigeration division.

Kerosene refrigerators were only built in Australia for the Australian market.

Advertising for the electric version, 1954. TC

NEW DESIGNS Set the pace

International Harvester Products
reach a *NEW HIGH* in performance

TODAY mechanized agriculture is becoming more widespread and more essential throughout the world. International Harvester in Australia, with its modern manufacturing facilities at Dandenong and Geelong and backed by 120 years of overseas manufacturing experience, is continually giving Australia's farmers new machines of new design and higher performance for greater farming efficiency. Recent new products include the AOS-6 tractor with matched 3-point linkage and toolbar equipment . . . the Super AW-6 tractor with increased horse power and hydraulic remote control for easier operation of matched implements . . . the new "Stamina-red" (engineered for stamina) Australian made AB-line of International Trucks and the streamlined 7-cubic-foot Defender "Country Style" refrigerator with specialized storage. A vast network of International Harvester Dealers sell and service these and other IH products. Your IH dealer, by serving agriculture through you, makes a working contribution to the economic security and well-being of Australia. International Harvester Company of Australia Pty. Ltd. and its Vice District Sales Offices in all Capital Cities. Works: Dandenong and Geelong, Victoria.

International Harvester

McCORMICK INTERNATIONAL TRACTORS AND FARM EQUIPMENT • INTERNATIONAL TRUCKS • DEFENDER REFRIGERATORS • INTERNATIONAL INDUSTRIAL POWER

BUILDER OF ESSENTIAL EQUIPMENT   FOR ESSENTIAL WORK

AD NO. 1411-TP-Full Page (11 x 9 inches)-4 colours
International Harvester Company of Australia Pty. Ltd.
For appearance in "Australia To-Day 1954" (Annual);
"Shipping, Commerce and Aviation of Australia" 1954 (Annual);
"Australasian Manufacturer" 1954 (Annual).
Prepared by Advertising Department, Merchandising Services, Melbourne.

Advertising for the 1954 Annual publications of "Australia To-Day"; "Shipping, Commerce and Aviation of Australia"; and Australasian Manufacturer" TC

1956 advertisement for the electric refrigerator. TC

DEFENDER
KEROSENE OPERATED REFRIGERATOR
PACKING SLIP

Serial No. *2901771*

LOOSE EQUIPMENT
Packed with Cabinet
1 Wire shelf with basket
3 Wire shelves, plain
1 Drip tray
5 Ice trays with dividers
2 Utility baskets
1 Hydrator with lid
1 Flue cap
1 Baffle
1 Support wire
2 Lamp glasses
1 Spare lamp wick
1 Ventilator with screws (fixed beneath cabinet)
1 Kero. burner with wick
1 Fuel tank
1 Flue brush (tied at back of cabinet)
1 Instruction book (in envelope)
1 Recipe book (in envelope)

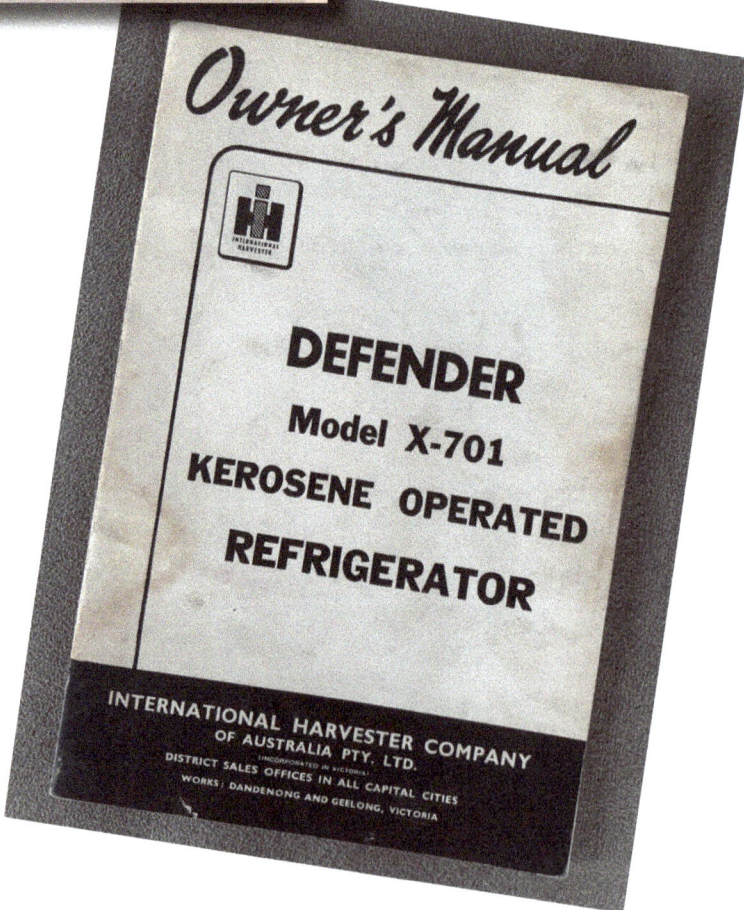

PACKED BY:

INSTRUCTION MANUAL
FOR
DEFENDER
KEROSENE-OPERATED
REFRIGERATOR
(MODEL X552)

DEFENDER

INTERNATIONAL HARVESTER COMPANY
OF AUSTRALIA PTY. LTD.
(INCORPORATED IN VICTORIA)
BRISBANE - SYDNEY - MELBOURNE - ADELAIDE - PERTH

RF.1   26-4-50

Owner's Manual

DEFENDER
Model X-701
KEROSENE OPERATED
REFRIGERATOR

INTERNATIONAL HARVESTER COMPANY
OF AUSTRALIA PTY. LTD.
(INCORPORATED IN VICTORIA)
DISTRICT SALES OFFICES IN ALL CAPITAL CITIES
WORKS: DANDENONG AND GEELONG, VICTORIA

Concepts for 1956
Refrigerator and Air Conditioners

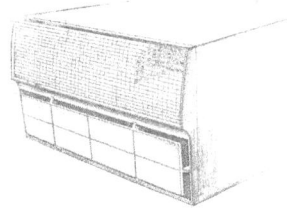

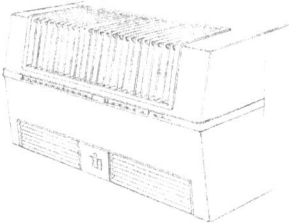

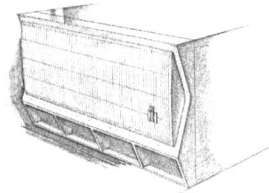

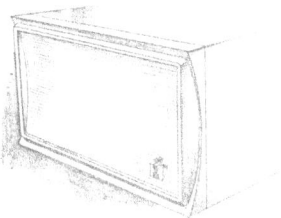

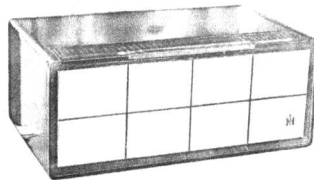

The following pages showcase some of the design process for styling the next series of refrigeration. Concept drawings and artwork begins two to three years in advance, the artist has to be able to draw for the future, getting the details correct for consumer demand.

These drawings were courtesy of Gregg Montgomery, former
Industrial Design Manager at International Harvester.

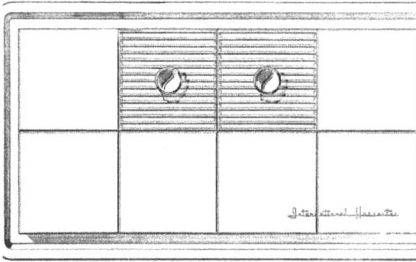

Which air conditioner would you like to see produced for 1956?

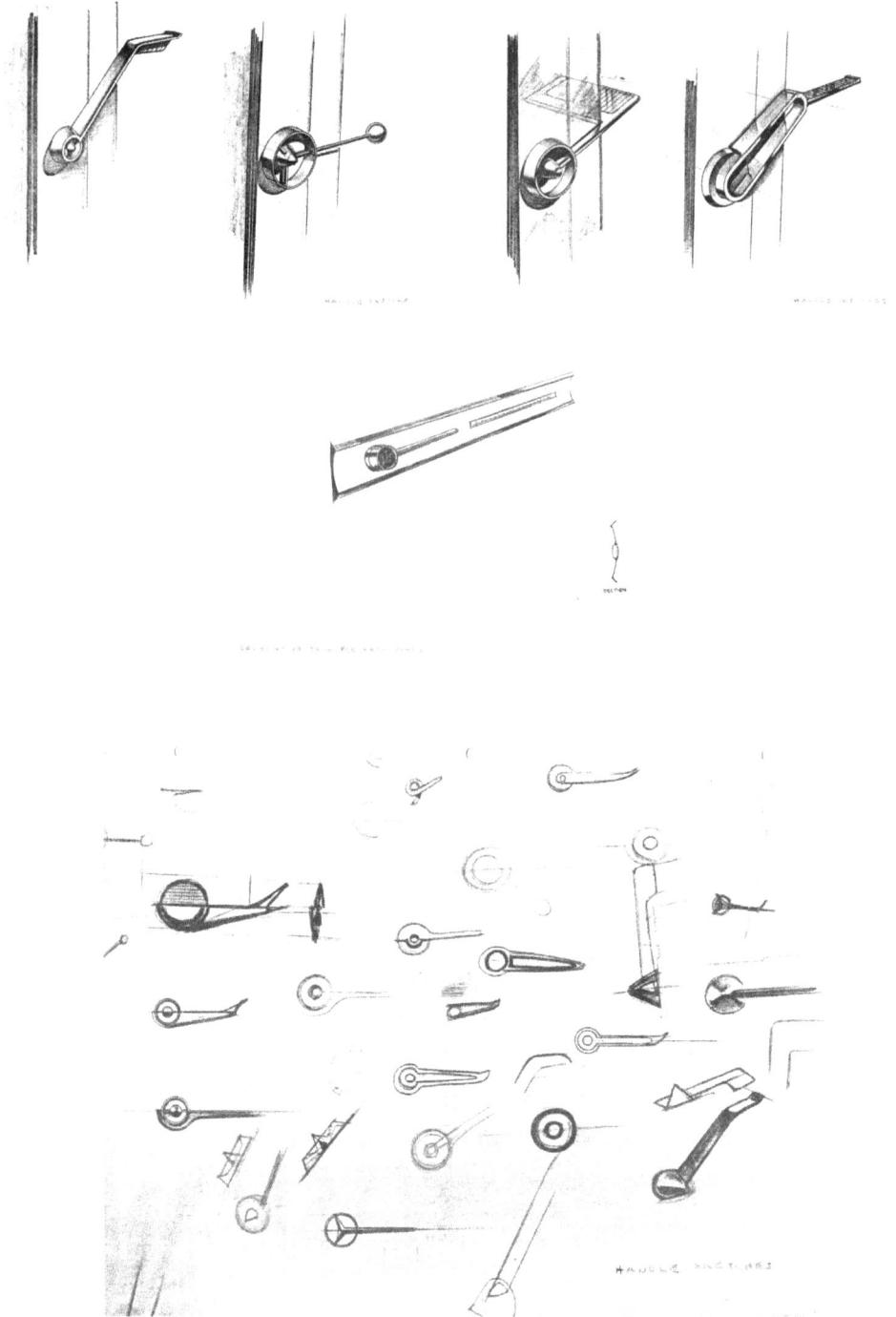

How to create a new handle. First, doodle and brainstorm, sketching a variety of ideas, realistic or otherwise. Second, use the elements from different drawings and apply finesse, drawing various options. Finally, present your drawings to the team and apply the final concept to modeling and add dimensions for production. Drawings provided by
Gregg Montgomery, Former Industrial Design Manager at International Harvester

Proposed design for the 1956 refrigerator series. Refrigerators have changed over the years, even in the short ten year time span that International Harvester built them.

Drawings provided with special thanks to Gregg Montgomery, former Industrial Design Manager at International Harvester.

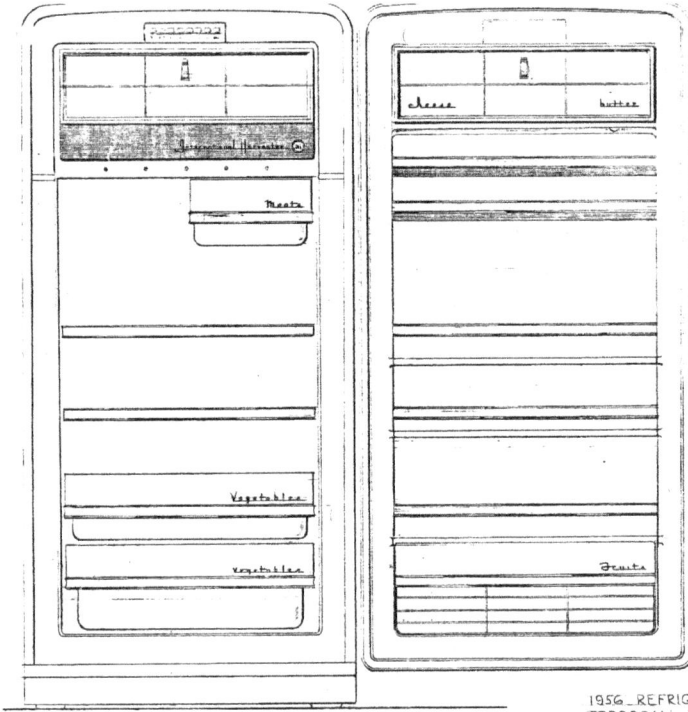

1956 REFRIGERATOR
PROPOSAL.

INTERNATIONAL HARVESTER COMPANY
INDUSTRIAL DESIGN DEPARTMENT
T. H. KOEBER   GENERAL OFFICE

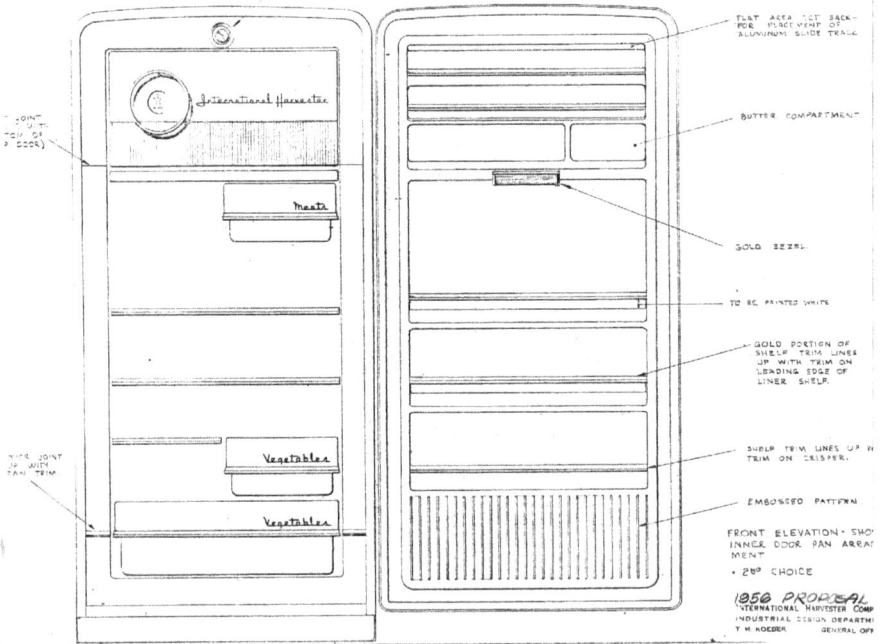

FLAT AREA SET BACK
FOR PLACEMENT OF
ALUMINUM SLIDE TRACK

BUTTER COMPARTMENT

GOLD BEZEL

TO BE PAINTED WHITE

GOLD PORTION OF
SHELF TRIM LINES
UP WITH TRIM ON
LEADING EDGE OF
LINER SHELF.

SHELF TRIM LINES UP W
TRIM ON CRISPER.

EMBOSSED PATTERN

FRONT ELEVATION - SHO
INNER DOOR PAN ARRAN
MENT

• 2ND CHOICE

1956 PROPOSAL
INTERNATIONAL HARVESTER COMP
INDUSTRIAL DESIGN DEPARTM
T. H. KOEBER   GENERAL OFF

The Following pages are a reference to the features and amount of charge needed for the various refrigerators and freezers.
Listed by first year built, early years units were built for multiple years.
Some details specific to serial numbers were not included for every change.

8H ORIGINAL RELEASE
SERIAL NUMBERS
MODEL 8H1   501 TO 10866   BUILT IN 1948
              10867 TO 47608   BUILT IN 1949

       8H3   501 TO 29111   BUILT IN 1948
              29112 TO 70466   BUILT IN 1949

       8H5   501 TO 29112   BUILT IN 1948
              29113 TO 55490   BUILT IN 1949

SERIAL NUMBERS FOR ALL MODELS WILL HAVE THE NUMBER FOLLOWED BY A LETTER AND A DIGIT. FOR EXAMPLE, SERIAL NUMBER FOR THE 8H3 IS 'H3-501-B8'
               MODEL: 8H3
               BUILD NUMBER: 501
               MONTH: B
               YEAR: 1948

CORRESPONDING MONTH TO THE FOLLOWING LETTERS: "I" WAS SKIPPED.
          JANUARY   A
          FEBRUARY   B
          MARCH   C
          APRIL   D
          MAY   E
          JUNE   F
          JULY   G
          AUGUST   H
          SEPTEMBER   J
          OCTOBER   K
          NOVEMBER   L
          DECEMBER   M

     8H1 STANDARD 8 CU FT; U-TYPE EVAPORATOR 8 CU FT 1/8 HP 115 VOLT
     8H3 DELUXE 8 CU FT; U-TYPE EVAPORATOR 8 CU FU 1/8 HP 115 VOLT
     8H5 SUPER DELUXE 8 CU FT; U-TYPE EVAPORATOR 8 CU FT 1/8 HP 115 VOLT

Series I Freezers

Year	1947	1946	1948	1948
Model	4 FC	**11 FC** s/n 504 to 9943	**11 FC** s/n 9944 Up	**15 FC**
Total Storage (Cu Ft)	4.2	11.1	11.1	15.8
Freezer storage (lbs of food)	150	385	385	553
Baskets (no.)	0	0	2	3
Dividers (no.)	0	0	2	4
Interior Light	No	No	No	Yes
Warning Alarm	No	No	Optional	Yes
Locking type handle	No	No	No	No
Temperature Control	Dial	Dial	Dial	Dial
Cabinet Width	33"	58"	58"	73 1/2"
Cabinet Depth	26"	29"	29"	29"
Cabinet Height	36 1/4"	37 1/4"	37 1/4"	37 1/4"
Weight	288.5	411	411	
Price	289.95			
Compressor Horsepower		1/4	1/4	
Voltage	115v	115v	115v	115v
Refrigerant Type	R-12	R-12	R-12	R-12
Original Refrigerant Charge in Ounces	12.5-13.0	15.0-17.0	15.0-17.0	18.5-19.5
Replacement Motor Charge in Ounces	14.0-14.5	16.5-18.5	16.5-18.5	20.0-21.0

New Filter Drier on both when recharging

Build Years for following Models:
4FC   1947-1949
11FC   1946-1949
15FC   1948-1949

SERIES II FREEZERS

Year	1950	1950	1950	1952
Model	**70**	**111**	**158**	**200**
Total Storage (Cu Ft)	7	11.1	15.8	20
Freezer storage (lbs of food)	245	388.5	553	700
Baskets (no.)	1	2	3	3
Dividers (no.)	1	2	2	2
Hostess Tray	No	No	No	Yes
Ice Cube Containers	No	No	No	Yes
Interior Light	No	Yes	Yes	Yes
Warning Alarm	No	Optional	Yes	Yes
Ice Cube Rack, 4 Trays	No	No	No	Yes
Locking type handle	Optional	Yes	Yes	Yes
Temperature Gauge	No	External View	External View	External View
Temperature Control	Yes	Yes	Yes	Yes
Cabinet Width	44"	58"	73 1/2"	73 1/2"
Cabinet Depth	27"	29 1/4"	29 1/4"	29"
Cabinet Height	36"	37 1/4"	37 1/4"	36"
Weight	287	406	505	530
Price				
Compressor Horsepower	1/6	1/4	1/3	1/3
Voltage	115v	115v	115v	115v
Refrigerant Type	R-22	R-22	R-22	R-22
Original Refrigerant Charge in Ounces	10.0-11.5	12.0-13.0	15.5-17.0	15.5-17.0
Replacement Motor Charge in Ounces	11.5-13.0	13.5-14.5	17.0-18.5	17.0-18.5

New Filter Drier on both when

Build Years for following Models:

70   1950-1952

111   1950-1952

158   1950-1952

200   1952- Early 1953 (replaced with L-20)

L - SERIES FREEZERS

	Year	1953	1953	1953	1953	1953	1953
Model		**L-7**	**L-12**	**L-16**	**L-20**	**L-9-V**	**L-14-V**
Total Storage (Cu Ft)		7	11.1	16	20	8.7	14
Freezer storage (lbs of food)		245	388.5	560	700	305	490
Baskets (no.)		1	2	3	3	2	3
Dividers (no.)		1	2	2	2	N/A	N/A
Hostess Tray		No	Yes	Yes	Yes	N/A	N/A
Utility Shelf (no.)		None	2	3	4	N/A	N/A
Pantry-Dor		N/A	N/A	N/A	N/A	3	No
Ice Cube Containers		No	Yes	Yes	Yes	N/A	N/A
Interior Light		Yes	Yes	Yes	Yes	Yes	Yes
Warning Light		Yes	Yes	Yes	Yes	Yes	Yes
Juice Dispenser		N/A	N/A	N/A	N/A	0	3
Ice Cube Rack, 4 Trays		No	No	No	Yes	N/A	N/A
Locking type handle		Optional	Yes	Yes	Yes	No	Yes
Automatic Temperature Control		Yes	Yes	Yes	Yes	Yes	Yes
Cabinet Width		44"	58"	73 1/2"	73 1/2"	29 1/2"	29 3/4"
Cabinet Depth		27"	29"	29"	29"	29 3/8"	30 5/8"
Cabinet Height		36"	36"	36"	36"	59 5/8"	68 1/2"
Weight		287	406	505	530	310	375
Price		289.95	419.95	519.95	599.95		
Compressor Horsepower		1/6	1/4	1/3	1/3	1/3	1/3
Voltage		115v	115v	115v	115v	115v	115v
Refrigerant Type		R-22	R-22	R-22	R-22	R-22	R-22
Original Refrigerant Charge in Ounces		10.0-11.5	12.0-13.0	15.5-17.0	15.5-17.0	11.0	14.5-15.0
Replacement Motor Charge in Ounces		11.5-13.0	13.5-14.5	17.0-18.5	17.0-18.5	12.5	16.0-16.5

New Filter Drier on both when
    recharging

M - SERIES FREEZERS

Year	1954	1954	1954	1954	1954	1954	1954
Model	**M-7**	**M-12**	**M-16**	**M-20**	**MV-9**	**MV-15**	**MV-19**
Total Storage (Cu Ft)	7	12	16	20	8.7	14.7	18.9
Freezer storage (lbs of food)	245	420	560	700	305	514	665
Baskets (no.)	1	2	3	4	2	3	4
Dividers (no.)	1	2	3	2	N/A	N/A	N/A
Decorator Fabric	No	Yes	Yes	Yes	Yes	Yes	Yes
Pantry-Dor	N/A	N/A	N/A	N/A	3	4	4
Interior Light	Yes	Yes	Yes	Yes	Yes	Yes	Yes
Warning Light	Yes	Yes	Yes	Yes	Yes	Yes	Yes
Juice Dispenser	0	2	2	2	0	0	3
Locking type handle	Optional	Yes	Yes	Yes	No	Yes	Yes
Automatic Temperature Control	Yes	Yes	Yes	Yes	Yes	Yes	Yes
Cabinet Width	44"	52 1/4"	65 3/8"	65 3/8"	29 1/2"	29 1/2"	36 1/2"
Cabinet Depth	27"	26 1/2"	26 1/2"	26 1/2"	29 3/8"	29 1/2"	30"
Cabinet Height	36"	37"	37"	37"	59 5/8"	68 1/2"	72 1/4"
Weight	264	366	437	460	310	390	451
Price							
Compressor Horsepower	1/6	1/4	1/3	1/3	1/3	1/3	1/3
Voltage	115v	115v	115v	115v	115v	115v	115v
Refrigerant Type	R-22	R-22	R-22	R-22	R-22	R-22	R-22
Original Refrigerant Charge in Ounces	10.0-11.5	13.0-13.5	14.5-15.0	16.0-16.5	11.0	14.5-15.0	14.5-15.0
Replacement Motor Charge in Ounces	11.5-13.0	14.5-15.0	16.0-16.5	17.5-18.0	12.5	16.0-16.5	16.0-16.5

New Filter Drier on both when recharging

A - Series Freezers

	Year	1955	1955	1955	1955	1955	1955
Model		**A-7**	**A-12**	**A-16**	**A-20**	**AV-15**	**AV-19**
Total Storage (Cu Ft)		7	12	16	20	14.7	18.9
Freezer storage (lbs of food)		245	420	560	700	514	665
Baskets (no.)		1	2	3	4	3	4
Dividers (no.)		1	2	2	3	N/A	N/A
Decorator Fabric		No	Yes	Yes	Yes	Yes	Yes
Pantry-Dor		N/A	N/A	N/A	N/A	3	3
Interior Light		Yes	Yes	Yes	Yes	Yes	Yes
Warning Light		Yes	Yes	Yes	Yes	Yes	Yes
Juice Dispenser (on door)		N/A	N/A	N/A	N/A	4-can wide	5-can wide
Locking type handle		Optional	Yes	Yes	Yes	Yes	Yes
Automatic Temperature Control		Yes	Yes	Yes	Yes	Yes	Yes
Cabinet Width		44"	52 1/4"	65 3/8"	65 3/8"	29 1/2"	36 1/2"
Cabinet Depth		27"	26 1/2"	26 1/2"	26 1/2"	29 1/2"	30"
Cabinet Height		36"	37"	37"	37"	68 1/2"	72 1/4"
Crated Weight		325	419	437	579	452	540
Price		269.95	369.95	469.95	529.95	499.95	579.95
Compressor Horsepower			1/4	1/3	1/3	1/3	1/3
Voltage		115v	115v	115v	115v	115v	115v
Refrigerant Type		R-22	R-22	R-22	R-22	R-22	R-22
Original Refrigerant Charge in Ounces		10.0-11.5	13.0-13.5	14.5-15.0	16.0-16.5	14.5-15.0	14.5-15.0
Replacement Motor Charge in Ounces		11.5-13.0	14.5-15.0	16.0-16.5	17.5-18.0	16.0-16.5	16.0-16.5

New Filter Drier on both when recharging

8H- SERIES REFRIGERATORS

Year	1948	1948	1948	1949
Model	**8H1** S/N 501 to 27967	**8H2**	**8H3**	**8H1** S/N 27968 Up
Net volume (Cu Ft)	8	8	8	8
Freezer storage (pounds	35	35	35	35
Ice cube trays (No. and type)	4, Aluminum Grid, plain	4, Aluminum, Tilt-out with Release Lever	4, Aluminum, Tilt-out with Release Lever	4, Aluminum Grid, plain
Shelf area (sq Ft)	13.3	13.4	16.3	13.3
Shelf Finish	Steel Lacquered Finish	Steel Lacquered Finish	Stainless Steel	Steel Lacquered Finish
Shelf Full	3	1	1	3
Shelf Half	0	1	2	0
Shelf, Small Folding Wire	0	1	2	0
Meat Drawer capacity (Cubic Inches)	320	780	780	320
Meat Drawer Material	Glass	Porcelian, Glass Cover	Porcelian, Glass Cover	Glass
Chill tray	N/A	N/A	N/A	N/A
Crisper (number)	0	1	2	1
Crisper Material	N/A	Porcelain Enameled Steel	Porcelain, Enameled Steel	Porcelain, Enameled Steel
Crisper Capacity (Quarts)	0	13	26	13
Crisper Location	N/A	Bottom Left	Bottom	Bottom Left
Pantry -Dor	N/A	N/A	N/A	N/A
Pantry Bin for Dry Storage	No	Yes, 2 cu.ft.	Yes, 2 cu.ft.	No
Bottle Opener Type	No	No	No	No
"Tele-Temp" Thermometer	No	No	Yes	No
Cabinet Width	31"	31"	31"	31"
Cabinet Depth	26"	26"	26"	26"
Cabinet Height	61 1/2"	61 1/2"	61 1/2"	61 1/2"
Weight	282	295	306	282
Compressor Horsepower				
Original Sales Price	$ 224.75	$ 259.75	$ 299.75	$ 219.75
Voltage	115v	115v	115v	115v
Refrigerant Type	R-12	R-12	R-12	R-12
Refrigerant Charge in Ounces	15.5 (+/- .5)	15.5 (+/- .5)	15.5 (+/- .5)	15.5 (+/- .5)

Build Years for following Models:
8H1   1948-1949
8H3   1948-1949
8H5   1948-1949

H AND U - SERIES REFRIGERATORS

Year	1950	1950	1950	1950	1950	1950
Model	**H-92**	**H-84**	**H-74**	**U-95**	**U-87**	**U-76**
Net volume (Cu Ft)	9.2	8.4	7.4	9.5	8.7	7.6
Freezer storage (lbs of food)	50	50	35	35	35	35
Ice cube trays (No. and type)	4 Lever Release	4 - 2 Lever Release, 2 Plastic Grid	2 Plastic Grid	4 2 Lever Release 2 Plastic Grid	3 Plastic Grid	2 Plastic Grid
Shelf area (sq Ft)	18.5	17.2	14.5	16.2	14.9	11.6
Shelf Finish	Stainless Steel	Chrome plated	Zinc Plated	Chrome Plated	Chrome Plated	Zinc Plated
Shelf Full	2	2	3	3	2	2
Shelf Half	1	1	0	0	0	0
Bottle Storage (quart bottles)	12	12		12	12	
Meat Drawer capacity (lbs)	10.8	10.8	14.5			8.6
Meat Drawer Material	Porcelain, Covered	Porcelain, Covered	Ice blue Plastic	Glass	Glass	Plastic, No cover
Chill tray	Yes	Yes	Yes	None	None	No
Crisper (number)	2	2	1	1	1	N/A
Crisper Material	Porcelain	Porcelain	Ice blue Plastic	Porcelain	Porcelain	N/A
Crisper Capacity (Qts of food)	23.3	26.1	11.3	14.2	14.5	N/A
Crisper Location	1 Across Bottom 1 Lower Left	1 Lower Left 1 Lower Right	Bottom Left	Across Bottom	Bottom Left	N/A
Pantry -Dor	4 Shelves	No	N/A	N/A	N/A	N/A
Bottle Opener Type	Regular	Regular	Regular	Regular	Regular	Regular
Egg-O-Mat	No	No	No	Yes	Yes	No
Pantry-Bin (unrefrigerated)	No	26 quarts	11.5 Quarts	No	No	
Cabinet Width	29 1/2"	29 1/2"	25"	29 1/2"	29 1/2"	25"
Cabinet Depth	29 1/8"	29 1/8"	28 7/8"	29 1/8"	29 1/8"	28 7/8"
Cabinet Height	59 5/8"	59 5/8"	54 3/8"	59 5/8"	59 5/8"	54 3/8"
Weight		307	249	292	292	249
Compressor Horsepower	1/8	1/8	1/9	1/8	1/8	1/9
Original Sales Price			$ 214.95	$ 259.95	$ 239.95	$ 199.95
Voltage	115v	115v	115v	115v	115v	115v
Refrigerant Type	R-12	R-12	R-12	R-12	R-12	R-12
Refrigerant Charge in Ounces	15.5 (+/- .5)	15.5 (+/- .5)	12.2 (+/- .5)	15.5 (+/- .5)	15.5 (+/- .5)	12.2 (+/- .5)

HA AND UA - SERIES REFRIGERATORS

Year	1951	1951	1951	1951	1951	1951	1951
Model	**HA-92**	**HA-84**	**HA-83**	**HA-82**	**HA-74**	**UA-95**	**UA-87**
Net volume (Cu Ft)	9.2	8.4	8.4	8.2	7.4	9.5	8.7
Freezer storage (lbs of food)	50	50	50	35	35	35	35
Ice cube trays (No. and type)	4 Lever Release	4 - 2 Lever Release, 2 Plastic Grid	3 Plastic Grid	2 Plastic Grid	2 Plastic Grid	4 2 Lever Release 2 Plastic Grid	3 Plastic Grid
Shelf area (sq Ft)	18	16.7	17.2	14.9	13.7	16.4	15.1
Shelf Finish	Stainless Steel	Chrome plated	Chrome Plated	Chrome Plated	Zinc Plated	Chrome Plated	Chrome Plated
Shelf Full	2	2	3	3	3	3	3
Shelf Half	1	0	1	0	0	0	0
Bottle Storage (quart bottles)	12	12	12	25	25	24	24
Meat Drawer capacity (lbs)	10.8	10.8	N/A	N/A	N/A	16.8	16.8
Meat Drawer Material	Porcelain, Covered	Porcelain, Covered	Plastic, No Cover	Plastic, No cover	Plastic, No Cover	Porcelain, Covered	Porcelain, Covered
Chill tray	Yes	Yes	Yes	Yes	Yes	None	None
Crisper (number)	2	2	1	1	None	1	1
Crisper Material	Porcelain	Porcelain	Porcelain	Porcelain	N/A	Porcelain	Porcelain
Crisper Capacity (Qts of food)	23.3	18.4	14.5	16	N/A	14.2	14.5
Crisper Location	1 Across Bottom 1 Lower Left	1 Lower Left 1 Lower Right	Lower Left	Across Bottom	N/A	Across Bottom	Lower Left
Pantry -Dor	4 Shelves	3 Shelves	N/A	N/A	N/A	3 Shelves	N/A
Butter Keeper	Yes	No	No	No	No	No	No
Bottle Opener Type	Magnetic	Magnetic	Regular	Regular	Regular	Regular	Regular
Egg-O-Mat	No	No	No	No	No	Yes	Yes
Color Inserts	Yes	Yes	Yes	Yes	No	Yes	Yes
Cabinet Width	29 1/2"	29 1/2"	29 1/2"	24 7/8"	24 7/8"	29 1/2"	29 1/2"
Cabinet Depth	29 1/8"	29 1/8"	29 1/8"	28 7/8"	28 7/8"	29 1/8"	29 1/8"
Cabinet Height	59 5/8"	59 5/8"	59 5/8"	54 3/8"	54 3/8"	59 5/8"	59 5/8"
Weight	302	296	291	233	220	292	292
Compressor Horsepower	1/8	1/8	1/8	1/9	1/9	1/8	1/8
Original Sales Price	$ 296.00	$ 264.00	$ 220.00	$ 200.00	$ 176.00	$ 240.00	$ 240.00
Voltage	115v	115v	115v	115v	115v	115v	115v
Refrigerant Type	R-12	R-12	R-12	R-12	R-12	R-12	R-12
Refrigerant Charge in Ounces	15.5 (+/- .5)	15.5 (+/- .5)	15.5 (+/- .5)	12.2 (+/- .5)	12.2 (+/- .5)	15.5 (+/- .5)	15.5 (+/- .5)

G - Series Refrigerators

Year	1952	1952	1952	1952	1952	1952	1952	1952
Model	**G-93D**	**G-85D**	**G-93**	**G-85**	**G-95**	**G-84**	**G-82**	**G-74**
Net volume (Cu Ft)	8.7	7.9	9.3	8.5	9.5	8.5	8.2	7.4
Freezer storage (lbs of food)	50	50	51	51	35	51	35	35
Ice cube trays (No. and type)	4 Lever Release	4 Lever Release	4 Lever Release	2 Lever Release, 2 Plastic Grid	2 Lever Release, 2 Plastic Grid	3 Plastic Grid	2 Plastic Grid	2 Plastic Grid
Shelf area (sq Ft)	16.3	15	18.3	17	16.4	17.5	14.9	13.7
Shelf Finish	Stainless Steel	Chrome Plated	Stainless Steel	Chrome Plated	Chrome Plated	Chrome Plated	Chrome Plated	Zinc
Shelf Full	2	2	2	2	3	2	3	3
Shelf Half	0	0	0	0	1	1	0	0
Bottle Storage (quart bottles)	12	12	12	12	24	12	25	25
Meat Drawer capacity (lbs)	10.8	10.8	10.8	10.8	16.8	19.6	15.5	14.5
Meat Drawer Material	Porcelain Enameled, Steel	Porcelain Enameled, Steel	Porcelain Enameled, Steel	Porcelain Enameled, Steel	Porcelain Enameled, Steel	N/A	N/A	N/A
Meat Drawer Cover	Plastic Translucent	Plastic Translucent	Plastic Translucent	Plastic Translucent	Plastic Translucent	N/A	N/A	N/A
Chill tray	Yes	Yes	Yes	Yes	No	Yes	Yes	Yes
Crisper (number)	2	2	2	2	1	1	1	No
Crisper Material	Porcelain Enameled, Steel	Porcelain Enameled, Steel	Porcelain Enameled, Steel	Porcelain Enameled, Steel	Porcelain Enameled, Steel	Porcelain Enameled, Steel	Porcelain Enameled, Steel	N/A
Crisper Capacity (Qts of food)	23.3	18.4	23.3	18.4	14.2	14.5	16	N/A
Crisper Location	1 across bottom, 1 Lower Left	1 Lower Left, 1 Lower Right	1 across bottom, 1 Lower Left	1 Lower Left, 1 Lower Right	Across Bottom	Lower Left	Across Bottom	N/A
Pantry -Dor	4 Shelf	3 Shelf	4 Shelf	3 shelf	3 shelf	2 Shelf	2 Shelf	N/A
Butter Keeper	Yes	Yes	Yes	Yes	Yes	No	No	No
Bottle Opener Type	Magnetic	Magnetic	Magnetic	Magnetic	Regular	Regular	Regular	Regular
Egg-O-Mat	No	No	No	No	Yes	No	No	No
Tri-Matic Defrost	Yes	Yes	No	No	No	No	No	No
Color Inserts	Yes	Yes	Yes	Yes	Yes	Yes	No	No
Spring Fresh Green Interior (Location)	Full Color	Breaker Strip, Door Panel	Breaker Strip, Door Panel	White Interior	White Interior	White Interior	Full Color	Green Plastic Freezer Door
Cabinet Width	29 1/2"	29 1/2"	29 1/2"	29 1/2"	29 1/2"	29 1/2"	24 7/8"	24 7/8"
Cabinet Depth	29 1/8"	29 1/8"	29 1/8"	29 1/8"	29 1/8"	29 1/8"	28 7/8"	28 7/8"
Cabinet Height	59 5/8"	59 5/8"	59 5/8"	59 5/8"	59 5/8"	59 5/8"	54 3/8"	54 3/8"
Weight	322	316	312	306	302	301	245	230
Compressor Horsepower	1/8	1/8	1/8	1/8	1/8	1/8	1/9	1/9
Original Sales Price								
Voltage	115v	115v	115v	115v	115v	115v	115v	115v
Refrigerant Type	R-12	R-12	R-12	R-12	R-12	R-12	R-12	R-12
Refrigerant Charge in Ounces	15.5 (+/- .5)	15.5 (+/- .5)	15.5 (+/- .5)	15.5 (+/- .5)	15.5 (+/- .5)	15.5 (+/- .5)	12.2 (+/- .5)	12.2 (+/- .5)

L - SERIES REFRIGERATORS

	L-105-DM / L-105-D (S/N 501 to 700)	L-105-DMS / L-105-DS (S/N 701 Up)	L-100-D (S/N 501 to 700)	L-100-DS (S/N 701 Up)	L-85-D	L-84-D / L-84-DM	L-84	L-82	L-74	L-103 (S/N 501 to 700)	L-103-S (S/N 701 Up)
Year	1953	1953	1953	1953	1953	1953	1953	1953	1953	1953	1953
Net volume (Cu Ft)	9.4	9.4	10	10	8.5	8.5	8.6	8.2	7.4	10.4	10.4
Freezer storage (lbs of food)	50	50	50	50	50	50	51	35	35	35	35
Ice cube trays (No. and type)	4 Lever Release	4 Lever Release	4 Lever Release	4 Lever Release	4 Lever Release	4 Lever Release	3 Plastic Grid	2 Plastic Grid	2 Plastic Grid	2 Lever Release, 2 Plastic Grid	2 Lever Release, 2 Plastic Grid
Shelf area (sq Ft)	19.6	19.5	19.6	19.6	17.3	17.3	17.6	14.9	13.7	17.1	16.4
Shelf Finish	Stainless Steel	Stainless Steel	Stainless Steel	Stainless Steel	Chrome Plated	Chrome Plated	Chrome Plated	Chrome Plated	Zinc	Chrome Plated	Chrome Plated
Shelf Full	2	2	3	3	2	2	2	3	3	3	3
Shelf Half	0	0	0	0	0	0	1	0	0	1	1
Bottle Storage (quart bottles)	12	12	up to 24	up to 24	12	12	12-24	25	25	24	24
Meat Drawer capacity (lbs)	10.8	10.8	10.8	10.8	10.8	10.8	19.6	15.5	14.5	16.8	16.8
Meat Drawer Material	Porcelain Enameled, Steel	Porcelain Enameled, Steel	Porcelain Enameled, Steel	Porcelain Enameled, Steel	Porcelain Enameled, Steel	Porcelain Enameled, Steel	Plastic	Plastic	Plastic	Porcelain Enameled, Steel	Porcelain Enameled, Steel
Meat Drawer Cover	Plastic Translucent	Plastic Translucent	Plastic Translucent	Plastic Translucent	Plastic Translucent	Plastic Translucent	N/A	N/A	N/A	Plastic Translucent	Plastic Translucent
Chill tray	Yes	Yes	Yes	Yes	Yes	Yes	Yes	Yes	Yes	No	No
Crisper (number)	2	2	1	1	2	1	1	1	No	1	1
Crisper Material	Porcelain Enameled, Steel	Porcelain Enameled, Steel	Porcelain Enameled, Steel	Porcelain Enameled, Steel	Porcelain Enameled, Steel	Porcelain Enameled, Steel	Porcelain Enameled, Steel	Porcelain Enameled, Steel		Porcelain Enameled, Steel	Porcelain Enameled, Steel
Crisper Capacity (Qts of food)	23.3	23.3	14.2	14.2	18.4	14.5	14.5	16	N/A	14.2	14.2
Crisper Location	1 across bottom, 1 Lower Left	1 across bottom, 1 Lower Left	Across Bottom	Across Bottom	Across Bottom	Lower Left	Lower Left	Across Bottom	N/A	Across Bottom	Across Bottom
Pantry -Dor	5.5	4 shelf	5.5 Shelf	4 Shelf	3 shelf	2 Shelf	2 Shelf	2 Shelf	N/A	Yes	Yes
Butter Keeper	Yes	Yes	Yes	Yes	Yes	Yes	No	Yes	No	Yes	Yes
Bottle Opener Type	Magnetic	Magnetic	Magnetic	Magnetic	Regular	Magnetic	Regular	Regular	Regular	Regular	Regular
Egg-O-Mat	No	No	No	No	No	No	No	No	No	No	No
Tri-Matic Defrost	Yes	Yes	Yes	Yes	Yes	Yes	No	No	No	No	No
Defrost Container	Yes	Yes	Yes	Yes	Yes	Yes	N/A	N/A	N/A	N/A	N/A
Spring Fresh Green Interior (Location)	Full Color	Full Color	Full Color	Full Color	Full Color	Breaker Strip, Door Panel	Breaker Strip, Door Panel	Full Color	Green Plastic Freezer Door, door panel, breaker strip	Breaker Strip, Door Panel	Breaker Strip, Door Panel
Cabinet Width	29 1/2"	29 1/2"	29 1/2"	29 1/2"	29 1/2"	29 1/2"	29 1/2"	24 7/8"	24 7/8"	29 1/2"	29 1/2"
Cabinet Depth	29 3/8"	29 3/8"	29 3/8"	29 3/8"	29 3/8"	29 3/8"	29 3/8"	28 1/2"	28 1/2"	29 3/8"	29 3/8"
Cabinet Height	59 3/4"	59 3/4"	59 3/4"	59 3/4"	59 3/4"	59 3/4"	59 3/4"	54 3/8"	54 3/8"	59 3/4"	59 3/4"
Weight	330	330	312	311	309	306	290	294	278	302	302
Compressor Horsepower	1/8	1/8	1/8	1/8	1/8	1/8	1/8	1/9	1/9	1/8	1/8
Original Sales Price	399.95	369.95	359.95	359.95	369.95	339.95	289.95	264.95	229.95	309.95	269.95
Voltage	115v	115v	115v	115v	115v	115v	115v	115v	115v	115v	115v
Refrigerant Type	R-12	R-12	R-12	R-12	R-12	R-12	R-12	R-12	R-12	R-12	R-12
Refrigerant Charge in Ounces	15.5 (+/-.5)	15.5 (+/-.5)	15.5 (+/-.5)	15.5 (+/-.5)	15.5 (+/-.5)	15.5 (+/-.5)	15.5 (+/-.5)	12.2 (+/-.5)	12.2 (+/-.5)	15.5 (+/-.5)	15.5 (+/-.5)

M - SERIES REFRIGERATORS

	M-105-DX	M-105-D	M-104	M-85-D, M-85-DL	M-85, M-85-L	M-82, M-82-L	M-75, M-75-L
Year	1954	1954	1954	1954	1954	1954	1954
Model	M-105-DX	M-105-D	M-104	M-85-D, M-85-DL	M-85, M-85-L	M-82, M-82-L	M-75, M-75-L
Left Hand Option	No	No	No	Yes	Yes	Yes	Yes
Net volume (Cu Ft)	10.2	10.2	10.4	8.3	8.4	8.2	7.4
Freezer storage (lbs of food)	50	50	35	50	50	35	35
Ice cube trays (No. and type)	4 Lever Release, Yellow Handle, Bronze Tray	4 Lever Release, Yellow Handle	4 Plastic Grid	4 Plastic Grid	3 Plastic Grid	2 Plastic Grid	2 Plastic Grid
Shelf area (sq Ft)	17.4	17.4	17.1	15.1	15.1	14.9	13.7
Shelf Finish	Stainless Steel	Stainless Steel	Chrome Plated	Chrome Plated	Chrome Plated	Chrome Plated	Chrome Plated
Shelf Full	2	3	3	2	2	3	3
Shelf Half	0	0	1	0	0	0	0
Meat Drawer capacity (lbs)	10.8	10.8	16.8	10.8	10.8	N/A	19.6
Meat Drawer Material	Porcelain Enameled, Steel	Porcelain Enameled, Steel	Porcelain Enameled, Steel	Porcelain Enameled, Steel	Porcelain Enameled, Steel	N/A	Plastic (chill tray)
Meat Drawer Cover	Plastic Translucent	Plastic Translucent	Plastic Translucent	N/A	N/A	N/A	N/A
Chill tray	No	No	No	No	No	Yes	Yes
Crisper (number)	2	1	1	2	1	1	0
Crisper Material	Porcelain Enameled, Steel	Porcelain Enameled, Steel	Porcelain Enameled, Steel	Porcelain Enameled, Steel	Porcelain Enameled, Steel	Porcelain Enameled, Steel	N/A
Crisper Capacity (Qts of food)	23.3	14.2	14.2	18.4	14.5	16	N/A
Crisper Location	1 across bottom, 1 Lower Left	Across Bottom	Across Bottom	Across Bottom	Lower Left	Across Bottom	N/A
Pantry -Dor	7	7	6	3	2	2 Shelf	No
Butter Keeper	Yes	Yes	Yes	Yes	No	Yes	No
Bottle Opener Type	Magnetic	Magnetic	Regular	Magnetic	Regular	Regular	Regular
Glide-Out Shelf	1	No	No	No	No	No	No
Push-Button Defrost	Yes, with all weather control	Yes, with all weather control	No	Yes	No	No	No
Defrost Container	Yes	Yes	No	Yes	Yes	No	No
Decorator	Yes, with fabric	Yes	Yes	Yes	Yes	Yes	Yes
Cabinet Width	29"	29"	29"	29"	29"	24 3/8"	24 3/8"
Cabinet Depth	29"	29"	29"	29"	29"	28"	28"
Cabinet Height	59 1/2"	59 1/2"	59 1/2"	59 1/2"	59 1/2"	54 3/8"	54 3/8"
Weight							
Compressor Horsepower	1/8	1/8	1/8	1/8	1/8	1/9	1/9
Original Sales Price							
Voltage	115v	115v	115v	115v	115v	115v	115v
Refrigerant Type	R-12	R-12	R-12	R-12	R-12	R-12	R-12
Refrigerant Charge in Ounces	15.5 (+/- .5)	15.5 (+/- .5)	15.5 (+/- .5)	15.5 (+/- .5)	15.5 (+/- .5)	12.2 (+/- .5)	12.2 (+/- .5)

A - Series Refrigerators

Year	1955	1955	1955	1955		1955	1955
Model	**A-120-D** **A-120-DL**	**A-106-D** **A-106-DX**	**A-104**	**A-95-D**	**A-85-D,** **A-85-DL**	**A-85,** **A-85-L**	**A-75,** **A-75-L**
Left Hand Option	Yes	No	No	No	Yes	Yes	Yes
Net volume (Cu Ft)	12	10.1	10.6	9.4	8.5	8.5	7.4
Freezer storage (lbs of food)	65	50	35	50	35	35	35
Ice cube trays (No. and type)	4 Lever Release, Yellow Handle, Copper Tray	4 Lever Release, Yellow Handle	3 Plastic Grid	3 Plastic Grid	2 Plastic Grid	2 Plastic Grid	2 Plastic Grid
Shelf area (sq Ft)	19.6	17	16.4	14.8	14.8	14.8	13.7
Shelf Finish	Stainless Steel	Stainless Steel	Chrome Plated	Chrome Plated	Chrome Plated	Chrome Plated	Zinc Plated
Shelf Full	3	2	3	3	3	3	3
Meat Drawer capacity (lbs)	10.8	10.8	10.8	10.8	15; chill tray	15; chill tray	14.5; Chill tray
Meat Drawer Material	Porcelain Enameled, Steel	Porcelain Enameled, Steel	Porcelain Enameled, Steel	Porcelain Enameled, Steel	Plastic	Plastic	Plastic
Meat Drawer Cover	Defrost Tray	Defrost Tray		Defrost Tray	Defrost tray		
Chill tray	No	No	No	No	Yes	Yes	Yes
Crisper (number)	1	2	1	1	1	1	0
Crisper Material	Porcelain Enameled, Steel	Porcelain Enameled, Steel	Porcelain Enameled, Steel	Porcelain Enameled, Steel	Porcelain Enameled, Steel	Porcelain Enameled, Steel	Optional, special order
Crisper Capacity (Qts of food)	29.4	23	13.8	13.8	16	16	
Crisper Location	Across Bottom	1 across bottom, 1 Lower Left	Across Bottom	Across Bottom	Across Bottom	Across Bottom	
Pantry -Dor	7.5	7	6	4	4	4	No
Butter Keeper with Tray	One Pound	One Pound	One Pound	1/4 Pound	1/4 Pound	1/4 Pound	No
Bacon Tray	Yes	No	No	No	No	No	No
Bottle Opener Type	Magnetic	Magnetic	Regular	Regular	Regular	Regular	Regular
Glide-Out Shelf	2	1	No	No	No	No	No
Push Button Defrost	Yes	Yes		Yes	Yes		
Summer-Winter Control	Automatic	Automatic	No	Manual	Manual	Manual	Manual
Defrost Container	Yes	Yes	No	Yes	Yes	No	No
Decorator	Yes, with fabric	Yes, DX with fabric	Yes	Yes	Yes	Yes	Yes
Cabinet Width	29 1/2"	29 1/2"	29 1/2"	29 1/2"	25"	25"	25"
Cabinet Depth	28"	28"	28"	28"	28"	28"	28"
Cabinet Height	59 1/2"	59 1/2"	59 1/2"	59 1/2"	54 3/8"	54 3/8"	54 3/8"
Crated Weight	387	379	361	366		295	280
Compressor Horsepower	1/8	1/8	1/8	1/8	1/8	1/8	1/8
Original Sales Price	449.95	419.95	309.95			239.95	189.95
Voltage	115v	115v	115v	115v	115v	115v	115v
Refrigerant Type	R-12	R-12	R-12	R-12	R-12	R-12	R-12
Refrigerant Charge in Ounces	13.0 (+/- .5)	15.0 (+/- .5)	12.3 (+/- .5)	15.0 (+/- .5)	11.2 (+/-.5)	11.2 (+/-.5)	11.2 (+/-.5)
Pancake' type Compressors, 1/8 hp.		14.0 (+/-.5)	10.7 (+/-.5)	14.0 (+/-.5)	10.2 (+/-.5)	10.2 (+/-.5)	10.2 (+/-.5)
Serial Number for 'Pancake' Type compressor		A-106 #6854 up; A-106-DX #6102 up	#10957 and Up	A-95 501 up; A-95-D #12463 up	A-85-D #8011 up; A-85-DL #834 up	A-85 13985 up; A-85-L #1443 up	A-75 #11414 up; A-75-L #1125 up

REFRIGERATOR AND FREEZER NOTES:

YEAR: **MAKE:**   **MODEL:**   **SERIAL NUMBER:**